AF720656

Modern Trends in Structural and Solid Mechanics 3

Series Editor
Noël Challamel

Modern Trends in Structural and Solid Mechanics 3

Non-deterministic Mechanics

Edited by

Noël Challamel
Julius Kaplunov
Izuru Takewaki

iSTE

WILEY

First published 2021 in Great Britain and the United States by ISTE Ltd and John Wiley & Sons, Inc.

Apart from any fair dealing for the purposes of research or private study, or criticism or review, as permitted under the Copyright, Designs and Patents Act 1988, this publication may only be reproduced, stored or transmitted, in any form or by any means, with the prior permission in writing of the publishers, or in the case of reprographic reproduction in accordance with the terms and licenses issued by the CLA. Enquiries concerning reproduction outside these terms should be sent to the publishers at the undermentioned address:

ISTE Ltd
27-37 St George's Road
London SW19 4EU
UK

www.iste.co.uk

John Wiley & Sons, Inc.
111 River Street
Hoboken, NJ 07030
USA

www.wiley.com

© ISTE Ltd 2021
The rights of Noël Challamel, Julius Kaplunov and Izuru Takewaki to be identified as the authors of this work have been asserted by them in accordance with the Copyright, Designs and Patents Act 1988.

Library of Congress Control Number: 2020952868

British Library Cataloguing-in-Publication Data
A CIP record for this book is available from the British Library
ISBN 978-1-78630-718-7

MIX
Paper from responsible sources
FSC® C013604

Contents

Preface

Short Bibliographical Presentation of Prof. Isaac Elishakoff

This book is dedicated to Prof. Isaac Elishakoff by his colleagues, friends and former students, on the occasion of his seventy-fifth birthday.

Figure P.1. *Prof. Isaac Elishakoff*

For a color version of all the figures in this chapter, see www.iste.co.uk/challamel/mechanics3.zip.

Prof. Isaac Elishakoff is an international leading authority across a broad area of structural mechanics, including dynamics and stability, optimization and anti-optimization, probabilistic methods, analysis of structures with uncertainty, refined theories, functionally graded material structures, and nanostructures. He was born in Kutaisi, Republic of Georgia, on February 9, 1944.

Figure P.2. *Elishakoff in middle school in the city of Sukhumi, Georgia*

Elishakoff holds a PhD in Dynamics and Strength of Machines from the Power Engineering Institute and Technical University in Moscow, Russia (Figure P.4 depicts the PhD defense of Prof. Isaac Elishakoff).

Figure P.3. *Elishakoff just before acceptance to university. Photo taken in Sukhumi, Georgia*

Figure P.4. *Public PhD defense, Moscow Power Engineering Institute and State University; topic "Vibrational and Acoustical Fields in the Circular Cylindrical Shells Excited by Random Loadings", and dedicated to the evaluation of noise levels in TU-144 supersonic aircraft*

His supervisor was Prof. V. V. Bolotin (1926–2008), a member of the Russian Academy of Sciences (Figure P.5 shows Elishakoff with Bolotin some years later).

Figure P.5. *Elishakoff with Bolotin (middle), member of the Russian Academy of Sciences, and Prof. Yukweng (Mike) Lin (left), member of the US National Academy of Engineering. Photo taken at Florida Atlantic University during a visit from Bolotin*

Currently, Elishakoff is a Distinguished Research Professor in the Department of Ocean and Mechanical Engineering at Florida Atlantic University. Before joining the university, he taught for one year at Abkhazian University, Sukhumi in the

Republic of Georgia, and 18 years at the Technion – Israel Institute of Technology in Haifa, where he became the youngest full professor at the time of his promotion (Figure P.6 shows Elishakoff presenting a book to Prof. Josef Singer, Technion's former president).

Figure P.6. *Prof. Elishakoff presenting a book to Prof. J. Singer, Technion's President; right: Prof. A. Libai, Aerospace Engineering Department, Technion*

Elishakoff has lectured at about 200 meetings and seminars, including about 60 invited, plenary or keynote lectures, across Europe, North and South America, the Middle East and the Far East.

Prof. Elishakoff has made vital and outstanding contributions in a number of areas in structural mechanics. In particular, he has analyzed random vibrations of homogeneous and composite beams, plates and shells, with special emphasis on the effects of refinements in structural theories and cross-correlations. Free structural vibrations have been tackled using a non-trivial generalization of Bolotin's dynamic edge effect method. Nonlinear buckling has been investigated using a novel method, incorporating experimental analysis of imperfections. As a result, the fundamental concept of closing the gap – spanning the entire 20th century – between theory and practice in imperfection-sensitive structures has been proposed. Novel methods of evaluating structural reliability have been proposed, taking into account the error associated with various low-order approximations, as well as human error; innovative generalization of the stochastic linearization method has been advanced.

A non-probabilistic theory for treating uncertainty in structural mechanics has been established. Dynamic stability of elastic and viscoelastic structures with imperfections has been studied. An improved, non-perturbative stochastic finite element method for structures has been developed. The list of Elishakoff's remarkable research achievements goes on.

His research has been acknowledged by many awards and prizes. He is a member of the European Academy of Sciences and Arts, a Fellow of the American Academy of Mechanics and ASME, and a Foreign Member of the Georgian National Academy of Sciences. Elishakoff is also a recipient of the Bathsheva de Rothschild prize (1973) and the Worcester Reed Warner Medal of the American Society of Mechanical Engineers (2016).

Figure P.7. *Elishakoff having received the William B. Johnson Inter- Professional Founders Award*

Elishakoff is directly involved in numerous editorial activities. He serves as the book review editor of the "Journal of Shock and Vibration" and is currently, or has previously been an associate editor of the *International Journal of Mechanics of Machines and Structures, Applied Mechanics Reviews*, and *Chaos, Solitons & Fractals*. In addition, he is or has been on the editorial boards of numerous journals, for example *Journal of Sound and Vibration*, *International Journal of Structural Stability and Dynamics*, *International Applied Mechanics* and *Computers & Structures*. He also acts as a book series editor for Elsevier, Springer and Wiley.

Figure P.8. *Inauguration as the Frank Freimann Visiting Professor of Aerospace and Mechanical Engineering; left: Rev. Theodore M. Hesburgh, President of the University of Notre Dame; right: Prof. Timothy O'Meara, Provost*

Prof. Elishakoff has held prestigious visiting positions at top universities all over the world. Among them are Stanford University (S. P. Timoshenko Scholar); University of Notre Dame, USA (Frank M. Freimann Chair Professorship of Aerospace and Mechanical Engineering and Henry J. Massman, Jr. Chair Professorship of Civil Engineering); University of Palermo, Italy (Visiting Castigliano Distinguished Professor); Delft University of Technology, Netherlands (multiple appointments, including the W. T. Koiter Chair Professorship of the Mechanical Engineering Department – see Figure P.9); Universities of Tokyo and Kyoto, Japan (Fellow of the Japan Society for the Promotion of Science); Beijing University of Aeronautics and Astronautics, People's Republic of China (Visiting Eminent Scholar); Technion, Haifa, Israel (Visiting Distinguished Professor); University of Southampton, UK (Distinguished Visiting Fellow of the Royal Academy of Engineering).

Figure P.9. *Prof. Elishakoff with Prof. Warner Tjardus Koiter, Delft University of Technology (center), and Dr. V. Grishchak, of Ukraine (right)*

Figure P.10. *Elishakoff and his colleagues during the AIAA SDM Conference at Palm Springs, California in 2004; Standing, from right to left, are Prof. Elishakoff, the late Prof. Josef Singer and Dr. Giora Maymon of RAFAEL. Sitting is the late Prof. Avinoam Libai*

Elishakoff has made a substantial contribution to conference organization. In particular, he participated in the organization of the Euro-Mech Colloquium on “Refined Dynamical Theories of Beams, Plates and Shells, and Their Applications” in Kassel, Germany (1986); the *Second International Conference on Stochastic Structural Dynamics*, in Boca Raton, USA (1990); “International Conference on Uncertain Structures” in Miami, USA and Western Caribbean (1996). He also coordinated four special courses at the International Centre for Mechanical Sciences (CISM), in Udine, Italy (1997, 2001, 2005, 2011).

Prof. Elishakoff has published over 540 original papers in leading journals and conference proceedings. He championed authoring, co-authoring or editing of 31 influential and extremely well-received books and edited volumes.

Here follows some praise of his work and books:

– “It was not until 1979, when Elishakoff published his reliability study … that a method has been proposed, which made it possible to introduce the results of imperfection surveys … into the analysis …” (Prof. Johann Arbocz, Delft University of Technology, The Netherlands, *Zeitschrift für Flugwissenschaften und Weltraumforschung*).

– “He has achieved world renown … His research is characterized by its originality and a combination of mathematical maturity and physical understanding which is reminiscent of von Kármán …” (Prof. Charles W. Bert, University of Oklahoma).

– “It is clear that Elishakoff is a world leader in his field … His outstanding reputation is very well deserved …” (Prof. Bernard Budiansky, Harvard University).

– “Professor Isaac Elishakoff … is subject-wise very much an all-round vibrationalist” (P. E. Doak, Editor in Chief, *Journal of Sound and Vibration*, University of Southampton, UK).

– “This is a beautiful book …” (Dr. Stephen H. Crandall, Ford Professor of Engineering, M.I.T.).

– “Das Buch ist in seiner Aufmachunghervorragendgestaltet und kannalsäusserstwertvolleErganzung … wäzmstensempfohlenwerden …” [The book’s appearance is perfectly designed and can be highly recommended as a valuable addition.] (Prof. Horst Försching, Institute of Aeroelasticity, Federal Republic of Germany, *Zeitschrift für Flugwissenschaften und Weltraumforschung*).

– "Because of you, Notre Dame is an even better place, a more distinguished University" (Prof. Rev. Theodore M. Hesburgh, President, University of Notre Dame).

– "It is an impressive volume ..." (Prof. Warner T. Koiter, Delft University of Technology, The Netherlands).

– "This extremely well-written text, authored by one of the leaders in the field, incorporates many of these new applications ... Professor Elishakoff's techniques for developing the material are accomplished in a way that illustrates his deep insight into the topic as well as his expertise as an educator ... Clearly, the second half of the text provides the basis for an excellent graduate course in random vibrations and buckling ... Professor Elishakoff has presented us with an outstanding instrument for teaching" (Prof. Frank Kozin, Polytechnic Institute of New York, *American Institute of Aeronautics and Astronautics Journal*).

– "By far the best book on the market today ..." (Prof. Niels C. Lind, University of Waterloo, Canada).

– "The book develops a novel idea ... Elegant, exhaustive discussion ... The study can be an inspiration for further research, and provides excellent applications in design ..." (Prof. G. A. Nariboli, *Applied Mechanics Reviews*).

– "This volume is regarded as an advanced encyclopedia on random vibration and serves aeronautical, civil and mechanical engineers ..." (Prof. Rauf Ibrahim, Wayne State University, *Shock and Vibration Digest*).

– "The book deals with a fundamental problem in Applied Mechanics and in Engineering Sciences: How the uncertainties of the data of a problem influence its solution. The authors follow a novel approach for the treatment of these problems ... The book is written with clarity and contains original and important results for the engineering sciences ..." (Prof. P. D. Panagiotopoulos, University of Thessaloniki, Greece and University of Aachen, Germany, *SIAM Review*).

– "The content should be of great interest to all engineers involved with vibration problems, placing the book well and truly in the category of an essential reference book ..." (Prof. I. Pole, *Journal of the British Society for Strain Measurement*).

– "A good book; a different book ... It is hoped that the success of this book will encourage the author to provide a sequel in due course ..." (Prof. John D. Robson, University of Glasgow, Scotland, UK, *Journal of Sound and Vibration*).

– "The book certainly satisfies the need that now exists for a readable textbook and reference book …" (Prof. Masanobu Shinozuka, Columbia University).

– "[the] author ties together reliability, random vibration and random buckling … Well written … useful book …" (Dr. H. Saunders, *Shock and Vibration Digest*).

– "A very useful text that includes a broad spectrum of theory and application" (*Mechanical Vibration*, Prof. Haym Benaroya, Rutgers University).

– "A treatise on random vibration and buckling … The reviewer wishes to compliment the author for the completion of a difficult task in preparing this book on a subject matter, which is still developing on many fronts …" (Prof. James T. P. Yao, Texas A&M University, *Journal of Applied Mechanics*).

– "It seems to me a hard work with great result …" (Prof. Hans G. Natke, University of Hannover, Federal Republic of Germany).

– "The approach is novel and could dominate the future practice of engineering" (*The Structural Engineer*).

– "An excellent presentation … well written … all readers, students, and certainly reviewers should read this preface for its excellent presentation of the philosophy and raison d'être for this book. It is well written, with the material presented in an informational fashion as well as to raise questions related to unresolved … challenges; in the vernacular of film critics, 'thumbs up'" (Dr. R. L. Sierakowski, U.S. Air Force Research Laboratory, *AIAA Journal*).

– "This substantial and attractive volume is a well-organized and superbly written one that should be warmly welcomed by both theorists and practitioners … Prof. Elishakoff, Li, and Starnes, Jr. have given us a jewel of a book, one done with care and understanding of a complex and essential subject and one that seems to have ably filled a gap existing in the present-day literature and practice" (*Current Engineering Practice*).

– "Most of the subjects covered in this outstanding book have never been discussed exclusively in the existing treatises … (*Ocean Engineering*).

– "The treatment is scholarly, having about 900 items in the bibliography and additional contributors in the writing of almost every chapter … This reviewer believes that *Non-Classical Problems in the Theory of Elastic Stability* should be a useful reference for researchers, engineers, and graduate students in aeronautical, mechanical, civil, nuclear, and marine engineering, and in applied mechanics" (*Applied Mechanics Reviews*).

– "What more can be said about this monumental work, other than to express admiration? … The study is of great academic interest, and is clearly a labor of love. The author is to be congratulated on this work …" (Prof. H. D. Conway, Department of Theoretical and Applied Mechanics, Cornell University).

– "This book … is prepared by Isaac Elishakoff, one of the eminent solid mechanics experts of the 20th century and the present one, and his distinguished coauthors, will be of enormous use to researchers, graduate students and professionals in the fields of ocean, naval, aerospace and mechanical engineers as well as other fields" (Prof. Patricio A. A. Laura, Prof. Carlos A. Rossit, Prof. Diana V. Bambill, Universidad Nacional del Sur, Argentina, *Ocean Engineering*).

– "This book is an outstanding research monograph … extremely well written, informative, highly original … great scholarly contribution …. There is no comparable book discussing the combination of optimization and anti-optimization … magnificent monograph …. This book, which certainly is written with love and passion, is the first of its kind in applied mechanics literature, and has the potential of having a revolutionary impact on both uncertainty analysis and optimization" (Prof. Izuru Takewaki, Kyoto University, *Engineering Structures*).

– "This book is a collection of a surprisingly large number of closed form solutions, by the author and by others, involving the buckling of columns and beams, and the vibration of rods, beams and circular plates. The structures are, in general, inhomogeneous. Many solutions are published here for the first time. The text starts with an instructive review of direct, semi-inverse, and inverse eigenvalue problems. Unusual closed form solutions of column buckling are presented first, followed by closed form solutions of the vibrations of rods. Unusual closed form solutions for vibrating beams follow. The influence of boundary conditions on eigenvalues is discussed. An entire chapter is devoted to boundary conditions involving guided ends. Effects of axial loads and of elastic foundations are presented in two separate chapters. The closed form solutions of circular plates concentrate on axisymmetric vibrations. The scholarly effort that produced this book is remarkable" (Prof. Werner Soedel, then Editor-in-Chief of *Journal Sound and Vibration*).

– "The field has been brilliantly presented in book form …" (Prof. Luis A. Godoy *et al.*, Institute of Advanced Studies in Engineering and Technology, Science Research Council of Argentina and National University of Cordoba, Argentina, *Thin-Walled Structures*).

– "Elishakoff is one of the pioneers in the use of the probabilistic approach for studying imperfection-sensitive structures" (Prof. Chiara Bisagni and Dr. Michela Alfano, Delft University of Technology; *AIAA Journal*).

– "Recently, Elishakoff *et al.* presented an excellent literature review on the historical development of Timoshenko's beam theory" (Prof. Zhenlei Chen *et al.*, *Journal of Building Engineering*).

Professor Isaac Elishakoff is the author or co-author of an impressive list of seminal books in the field of deterministic and non-deterministic mechanics, presented below.

Books by Elishakoff

Ben-Haim, Y. and Elishakoff, I. (1990). *Convex Models of Uncertainty in Applied Mechanics.* Elsevier, Amsterdam.

Cederbaum, G., Elishakoff, I., Aboudi, J., Librescu, L. (n.d.). *Random Vibration and Reliability of Composite Structures.* Technomic, Lancaster.

Elishakoff, I. (1983). *Probabilistic Methods in the Theory of Structures.* Wiley, New York.

Elishakoff, I. (1999). *Probabilistic Theory of Structures.* Dover Publications, New York.

Elishakoff, I. (2004). *Safety Factors and Reliability: Friends or Foes?* Kluwer Academic Publishers, Dordrecht.

Elishakoff, I. (2005). *Eigenvalues of Inhomogeneous Structures: Unusual Closed-Form Solutions of Semi-Inverse Problems.* CRC Press, Boca Raton.

Elishakoff, I. (2014). *Resolution of the Twentieth Century Conundrum in Elastic Stability.* World Scientific/Imperial College Press, Singapore.

Elishakoff, I. (2017). *Probabilistic Methods in the Theory of Structures: Random Strength of Materials, Random Vibration, and Buckling.* World Scientific, Singapore.

Elishakoff, I. (2018). *Probabilistic Methods in the Theory of Structures: Solution Manual to Accompany Probabilistic Methods in the Theory of Structures: Problems with Complete, Worked Through Solutions.* World Scientific, Singapore.

Elishakoff, I. (2020). *Dramatic Effect of Cross-Correlations in Random Vibrations of Discrete Systems, Beams, Plates, and Shells.* Springer Nature, Switzerland.

Elishakoff, I. (2020). *Handbook on Timoshenko-Ehrenfest Beam and Uflyand-Mindlin Plate Theories.* World Scientific, Singapore.

Elishakoff, I. and Ohsaki, M. (2010). *Optimization and Anti-Optimization of Structures under Uncertainty.* Imperial College Press, London.

Elishakoff, I. and Ren, Y. (2003). *Finite Element Methods for Structures with Large Stochastic Variations*. Oxford University Press, Oxford.

Elishakoff, I., Lin, Y.K., Zhu, L.P. (1994). *Probabilistic and Convex Modeling of Acoustically Excited Structures*. Elsevier, Amsterdam.

Elishakoff, I., Li, Y., Starnes Jr., J.H. (2001). *Non-Classical Problems in the Theory of Elastic Stability*. Cambridge University Press, Cambridge.

Elishakoff, I., Pentaras, D., Dujat, K., Versaci, C., Muscolino, G., Storch, J., Bucas, S., Challamel, N., Natsuki, T., Zhang, Y., Ming Wang, C., Ghyselinck, G. (2012). *Carbon Nanotubes and Nano Sensors: Vibrations, Buckling, and Ballistic Impact*. ISTE Ltd, London, and John Wiley & Sons, New York.

Elishakoff, I., Pentaras, D., Gentilini, C., Cristina, G. (2015). *Mechanics of Functionally Graded Material Structures*. World Scientific/Imperial College Press, Singapore.

Books edited or co-edited by Elishakoff

Ariaratnam, S.T., Schuëller, G.I., Elishakoff, I. (1988). *Stochastic Structural Dynamics – Progress in Theory and Applications*. Elsevier, London.

Casciati, F., Elishakoff, I., Roberts, J.B. (1990). *Nonlinear Structural Systems under Random Conditions*. Elsevier, Amsterdam.

Chuh, M., Wolfe, H.F., Elishakoff, I. (1989). *Vibration and Behavior of Composite Structures*. ASME Press, New York.

David, H. and Elishakoff, I. (1990). *Impact and Buckling of Structures*. ASME Press, New York.

Elishakoff, I. (1999). *Whys and Hows in Uncertainty Modeling*. Springer, Vienna.

Elishakoff, I. (2007). *Mechanical Vibration: Where Do We Stand?* Springer, Vienna.

Elishakoff, I. and Horst, I. (1987). *Refined Dynamical Theories of Beams, Plates and Shells and Their Applications*. Springer Verlag, Berlin.

Elishakoff, I. and Lin, Y.K. (1991). *Stochastic Structural Dynamics 2 – New Applications*. Springer, Berlin.

Elishakoff, I. and Lyon, R.H. (1986), *Random Vibration-Status and Recent Developments*. Elsevier, Amsterdam.

Elishakoff, I. and Seyranian, A.P. (2002). *Modern Problems of Structural Stability*. Springer, Vienna.

Elishakoff, I. and Soize, C. (2012). *Non-Deterministic Mechanics*. Springer, Vienna.

Elishakoff, I., Arbocz, J., Babcock Jr., C.D., Libai, A. (1988). *Buckling of Structures: Theory and Experiment*. Elsevier, Amsterdam.

Lin, Y.K. and Elishakoff, I. (1991). *Stochastic Structural Dynamics 1 – New Theoretical Developments*. Springer, Berlin.

Noor, A.K., Elishakoff, I., Hulbert, G. (1990). *Symbolic Computations and Their Impact on Mechanics*. ASME Press, New York.

Figure P.11. *Elishakoff with his wife, Esther Elisha, M.D., during an ASME awards ceremony*

On behalf of all the authors of this book, including those friends who were unable to contribute, we wish Prof. Isaac Elishakoff many more decades of fruitful works and collaborations for the benefit of world mechanics, in particular.

Modern Trends in Structural and Solid Mechanics 1 – the first of three separate volumes that comprise this book – presents recent developments and research discoveries in structural and solid mechanics, with a focus on the statics and stability of solid and structural members.

The book is centered around theoretical analysis and numerical phenomena and has broad scope, covering topics such as: buckling of discrete systems (elastic chains, lattices with short and long range interactions, and discrete arches), buckling of continuous structural elements including beams, arches and plates, static investigation of composite plates, exact solutions of plate problems, elastic and inelastic buckling, dynamic buckling under impulsive loading, buckling and post-buckling investigations, buckling of conservative and non-conservative systems, buckling of micro and macro-systems. The engineering applications concern both small-scale phenomena with micro and nano-buckling up to large-scale structures, including the buckling of drillstring systems.

Each of the three volumes is intended for graduate students and researchers in the field of theoretical and applied mechanics.

Prof. Noël CHALLAMEL
Lorient, France
Prof. Julius KAPLUNOV
Keele, UK
Prof. Izuru TAKEWAKI
Kyoto, Japan
February 2021

1

Optimization in Mitochondrial Energetic Pathways

1.1. Optimization in neural and cell biology

Neuronal networks and their countless pathways, as well as their companion cells, the glia, are the foundation of our functioning brain. These neurons and glia send signals to each other and process information in tremendously complex ways, which we are only beginning to have some understanding of. How pathway choices in neurons lead to physiological or pathological responses are of critical interest. What causes one progression path to be followed rather than another? Are there underlying principles, for example, an optimization of energy or time, a minimization of materials or some other underlying rules and characteristics?

Each neuron cell is composed of numerous organelles and other components, each with their own function, all of which communicate intracellularly and respond as a team to the needs of the cell, as well as to extracellular signals. One of those organelles is the mitochondria. It is well established that the connections between energetics and mitochondria – the organelle responsible for almost all of the energy production in the cell – determine whether physiological or pathological pathways are taken at all levels: subcellular, cellular, tissue and organism. Dysfunction in the mechanisms of energy production appears to be at the center of neurological and neuropsychiatric pathologies. Thus, the profound interest is in understanding how these organelles function and govern their operations. One example of dysfunction is how secondary pathologies in traumatic brain injuries result from energetic dysfunction.

Chapter written by Haym BENAROYA.

Neurons and glia are the components of brain function, but energy homeostasis must be maintained in order to assure proper functioning, and begins within the subcellular organelle, the mitochondria. This homeostasis is the product of metabolic reactions that are coupled to energy demands in space and time throughout the brain, and are regulated by feedforward and feedback mechanisms. Any mismatch between supply and demand over significant time intervals invariably initiates cascades of dysfunction leading to well-known neurodegenerative and neuropsychiatric pathologies.

These mechanisms, at all scales, “choose” from numerous progression paths, some of which lead to dysfunction due, in large part, to ineffective energy production. Understanding how these “choices” are made requires us to formulate models of the mechanisms alluded to above. If there are governing optimal “choices” or mechanisms, then dysfunction and defects may also sometimes be optimal choices for the organism, and perhaps the optimizations are energy-dependent given the criticality of energy production and usage. Perhaps, optimal choices can lead to negative outcomes. Given the multiple constraints for a successful living organism, there may only be local sub-optimizations. Thus, when we refer to optimization, we are having the above discussion, about how we frame the multitude of progression paths within the cell and external to it. Evolution governed which organisms survived, and which did not, based on their fitness for the environment. At the cellular level, this may entail a minimization of energy use, or perhaps the quickest transfer of information between two neurons.

Understanding these optimality decisions can provide clues for clinical interventions and eventually, cures for some of humanity’s most serious neurodegenerative diseases. Optimal pathways may be identified via the multitude of techniques that have been developed for the physical sciences and engineering, taking the morphological (and mechanical), biochemical and metabolic constraints into account. Constraints such as signaling mechanisms at all scales, cell and organelle morphology, feedback mechanisms, and imbalances of energetics and other intermediate products of mitochondrial functioning are part of a possible formulation. The community is at the beginning of formulating such models. This overview aims to pull together a very brief summary of current thoughts and evidence that, at least for the mitochondrial organelle at the subcellular level, the responses are evolutionarily conserved local and global optimizations.

In the following sections, we discuss some of the key functions of the mitochondria and identify, or suggest how these appear to be evolutionarily conserved properties that are based on an optimization. While optimization in biology has been discussed for decades, the application of optimization methodologies to such systems has been slow to develop, in part due to an incomplete understanding of significant aspects of functioning and an incomplete dataset.

We also refer to the work of Elishakoff (e.g. 1994, 2003, and with Qiu 2001). Elishakoff developed the concept of anti-optimization, where system uncertainties are studied by combining conventional optimization methods with interval analysis. In this approach, the optimal solution is a domain, rather than a point and is a two-level process. At one level, the optimal values of system parameters are obtained, and at the other level, uncertainties are anti-optimized. The anti-optimization yields the least favorable and most favorable system response and relies on knowledge of the bounds of uncertainty, rather than probability density functions. Such an approach can be potentially useful in biological systems where data can be sparse, with uncertainties only known via the upper and lower bounds.

1.2. Mitochondria

Our brief overview is about the very exciting area of the intense biological research of the mitochondria, an intracellular organelle. The mitochondria's primary functions include the maintenance of energy homeostasis, cell integrity and survival (Simcox and Reeve 2016). The mitochondria variably comprise between about 20% of the cell, up to most of the cell volume, dynamically depending on the energetic needs of the cell. In the aggregate, mitochondria account for about 10% of body weight, attesting to their importance. These organelles produce up to 95% of a eukaryotic cell's energy through oxidative phosphorylation (Tzameli 2012), driven by an electrochemical proton gradient created by the respiratory chain housed within the mitochondria's inner membrane. Oxidative phosphorylation is the metabolic pathway in the mitochondrial matrix containing the cristae, where enzymes oxidize nutrients. Energy is released, producing adenosine triphosphate (ATP), a complex organic chemical that provides energy for many cellular processes. The human body consumes, on average, a quantity of ATP per day that approximates its body weight (Zick and Reichert 2011).

Eukaryotes are organisms with cells that have a nucleus enclosed within membranes, unlike prokaryotes (bacteria and archaea). Eukaryotes may also be multicellular and consist of many cell types.

The cristae are tight folds of the inner membrane studded with proteins, with the folds providing a significant increase in surface area (much in the same way as the folded cerebral cortex), over which the above energy-producing processes can occur. Cristae biogenesis, regulated through the large enzyme ATP synthase, closely links mitochondrial morphology to energy demand (Simcox and Reeve 2016).

Mitochondrial functions also include quality control through fission (division) and fusion (merging), iron–sulfur cluster formation, calcium handling, cell signaling, cell repair and maintenance, and reactive oxygen species (ROS) production (Simcox and Reeve 2016). ROS emission is harmful to the mitochondrial DNA, which is a byproduct of energy production (Kembro *et al.* 2013) and is at least partially eliminated. As part of their metabolic functions, mitochondria also perform critical programmed cell death functions called apoptosis (Vakifahmetoglu-Norberg *et al.* 2017), which are related to the function of fission. Figure 1.1 summarizes some of these functions.

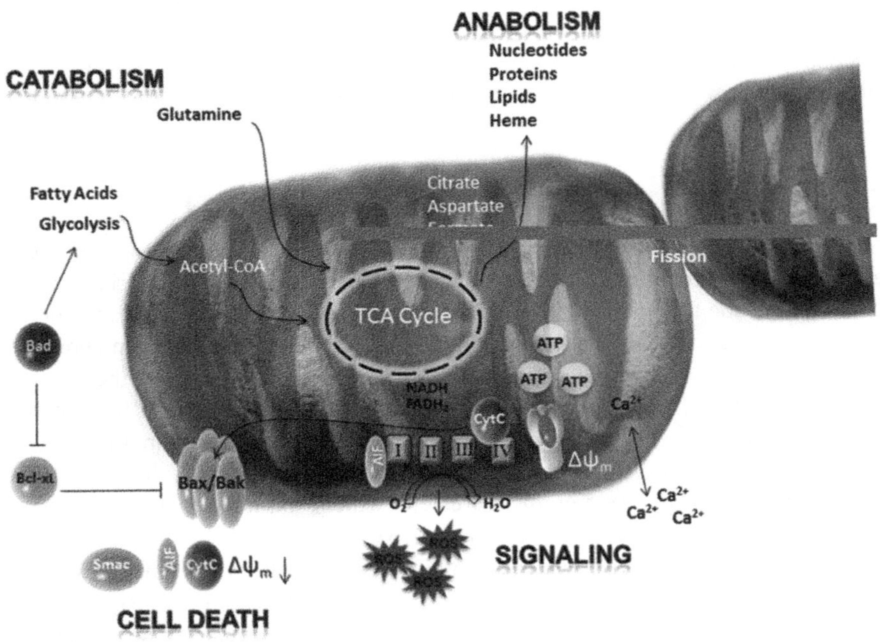

Figure 1.1. *Mitochondria shown undergoing fission/fusion. The respiratory complexes are shown, identified using roman numerals. Various mechanisms are also shown, including signaling, Ca^{2+} transport across the membrane, and others outside of our current scope (Vakifahmetoglu-Norberg et al. 2017, with permission). For a color version of this figure, see www.iste.co.uk/challamel/mechanics3.zip*

Mitochondria are believed to be the reason why complex cellular beings evolved from single celled entities. They originated as individual cell bacteria, but eventually integrated with our ancestral cells, leading to the current eukaryotic cells with nuclei. This ancestry partially explains why mitochondria, to this day, contain mtDNA, remnants of their own DNA.

The goal of this chapter is to refer to aspects of mitochondrial behavior that appear to be governed by optimization principles, which are constrained by biological limitations. It is reasonable to assume that the mathematical machinery of optimization can be useful in modeling some of these processes. Also, given that there are numerous mechanistic determinants of cell and intracellular functioning, we believe that these can be modeled using some of the principles of mechanics that govern engineered systems, as well as the frequently observed feedback and feedforward mechanisms that coordinate the multitude of processes within cells. Of course, biological systems are significantly more complex and require considerably more experimentation in order to ascertain behavior. We suggest the ideas herein with humility. A more detailed review of some of these ideas is available (Benaroya 2020), where the links between mitochondria, cellular energy production and the coupling of these to diseases and the body's response to injury are discussed.

1.3. General morphology; fission and fusion

Mitochondria exist in varying numbers, dependent on cell type, and sometimes form intracellular networks of interconnecting organelles called a reticulum, extending throughout the cytosol and in close contact with the nucleus, the endoplasmic reticulum, the Golgi network and the cytoskeleton (Benard *et al.* 2007). The cytosol is the intracellular fluid that surrounds all the organelles and other components of the cell. The endoplasmic reticulum, an organelle, has protein-related functions, and works with the Golgi network to deliver proteins to where they are needed within the cell.

The mitochondrial network is a net-like formation. "Mitochondrion" traces back to the Greek word meaning "thread grain". Mitochondria may exist as small isolated particles, or as extended filaments, networks or clusters connected via intermitochondrial junctions. Serial-section three-dimensional images showed filamentous mitochondria frequently linked into networks (Skulachev 2001).

Extended mitochondria, and electrically coupled mitochondrial clusters, can transmit power in the form of membrane potential between remote parts of the cell. The coupled clusters are switched on and off as needed, in order to avoid local damage due to a simultaneous discharge of too many of the organelles at once. As energy demand increases, isolated mitochondria unite into extended mitochondrial systems. In tissues composed of large cells with high energy demands, such as the brain, heart and kidneys, the so-called mitochondrial reticulum occupies much of the cell volume.

Certain physiological and pathological conditions lead to the decoupling of mitochondrial filaments and networks into single mitochondria. Extended mitochondrial systems of various topologies exist only when their energy coupling and transmitting machineries are functioning normally (Skulachev 2001). Muscle fibers require a specialized spatial organization of the mitochondrial network (Vinogradskaya *et al.* 2014).

Mitochondrial fission, fusion, motility and tethering, the four conserved and interdependent mitochondrial activities, alter the mitochondrial network shape and its distribution in the cell. A dysfunction of one activity can have consequences on another. For example, it is observed that the attenuation of fission disrupts the transport of mitochondria to neuronal synapses, resulting in detrimental effects on cell function. Tethering defects can reduce fission rates (Lackner 2014). Relative rates of mitochondrial fission and fusion govern the connectivity of the network, where energetic needs coordinate the two processes. Feedforward and feedback mechanisms coordinate the complex relationships between energy supply and demand. We anticipate that these activities operate within an optimal domain.

In complex polarized cells such as neurons, mitochondria must be actively transported and tethered to, and maintained in, active synaptic regions. Tethers are important for positioning mitochondria within the overall cell structure and also relative to other organelles.

Mitochondrial morphology (network organization) and bioenergetic functions are coupled bidirectionally (Benard *et al.* 2007). The metabolic needs of the cell optimize the organelle's bioenergetic capacity using frequent cycles of fusion and fission to adapt the morphology of the mitochondrial compartment to current supply and demand, as well as other required functions, of which there are many (Pagliuso *et al.* 2018). A disruption of these, for example, unopposed fission or fusion, adversely impacts cellular and organismal metabolism, leading to potentially devastating dysfunction (Wai and Langer 2016).

Fusion engages the entire mitochondrial compartment in respiratory active cells, maximizing ATP synthesis by mixing the matrix and the inner membrane, allowing close cooperation within the respiratory machinery. Metabolites, enzymes and mitochondrial gene products are spread throughout the mitochondrial compartment. Extensive adaptations of mitochondria to bioenergetic conditions occur at the level of the inner membrane ultrastructure and the remodeling of mitochondria cristae. A sudden need for metabolic energy results in cell stress and can lead to the formation of hyperfused mitochondrial networks. Such short-term stress exposure in starvation results in fusion that optimizes mitochondrial function and plays a beneficial role for the long-term maintenance of bioenergetic capacity. In a complementary way, irreversibly damaged mitochondria are eliminated by fission (autophagy), contributing to the maintenance of bioenergetic capacity (Westermann 2012). It is interesting, as a practical matter, that stresses to the cells benefit mitochondria in the same way as exercising in the gym benefits our muscles. Whether keeping the body at a proper range of temperatures when it is too cold, keeping up energy production during reduced caloric intake or matching energy needs during strenuous activity, the mitochondria comes out stronger and healthier, with larger numbers. Stress is an optimization mechanism.

Extensive disturbances to the dynamic balance between fission and fusion are linked to neurodegenerative and metabolic diseases (Chauhan *et al.* 2014). One purpose of this cycle between fission and fusion is to minimize the accumulation of reactive oxygen species (ROS). ROS are formed as a natural byproduct of the normal metabolism of oxygen and have important roles in cell signaling and homeostasis. However, during times of stress, ROS levels can increase dramatically, resulting in significant damage to cell structures. Cumulatively, this is known as oxidative stress.

Mitochondrial performance can be estimated by its bioenergetic capacity (ATP generation), metabolic capacity (mTOR activity) and damage accumulation (ROS production and/or mutation accumulation in mtDNA). Several nutrient sensing pathways link glucose metabolism to mitochondrial ATP, mTOR and ROS levels, which, in turn, directly or indirectly control proteins of fission–fusion machinery, like the fission proteins Drp1 and Fis1. The system flow chart is shown in Figure 1.2.

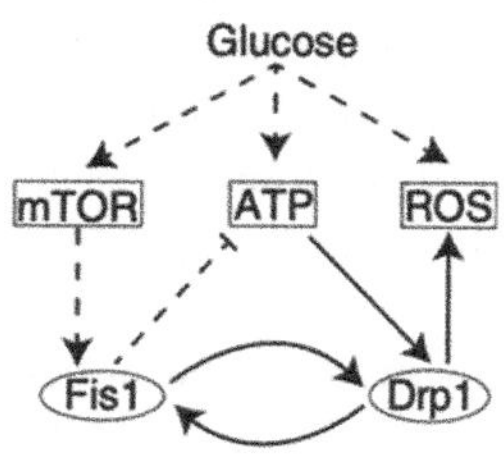

Figure 1.2. *Derivation of mitochondrial performance phenomenologically (Chauhan* et al. *(2014), with permission)*

The ATP module in the network above can, for example, be modeled by the following three first-order ordinary differential rate equations (Chauhan *et al.* 2014):

$$\frac{d[ATP]}{dt} = (J_{Glyc} + J_{Resp}) \cdot q_1 \cdot [Fis1] - J_{ATPconsume}$$
$$\frac{d[Drp1]}{dt} = k_1 \cdot [Fis1] - d_1 \cdot [drp1]$$
$$\frac{d[Fis1]}{dt} = k_2 \cdot [Drp1] + d_2 \cdot [Fis1]$$

[1.1]

$$J_{Glyc} = p_1 \left(\frac{[Glu]}{p_2}\right)^{p_3} \cdot \left(\frac{[Ca_m]^{2+}}{p_4 + [Ca_m]^{2+}}\right)$$
$$J_{Resp} = \left(\frac{p_5[ADP]}{[ADP] + p_6[ATP]}\right) \cdot \left(\frac{(\frac{[ATP]}{[ADP]})_m^{p_7}}{p_8^{p_9} + (\frac{[ATP]}{[ADP]})_m^{p_9}}\right) \cdot \left(1 - \frac{1}{e^{[Ca_m]^{2+}/p_{10}}}\right)$$
$$J_{cons} = (p_{11} + p_{12}[Ca_c]^{2+}) \cdot [ATP]$$

The quantities in the square brackets represent concentrations. J_{Glyc}, J_{Resp} and J_{cons} are the respective fluxes of glycolysis, mitochondrial respiration and ATP consumption in other cellular processes, derived phenomenologically. q_1 is the fission rate, k_1 and k_2 are the respective synthesis rates, and d_1 and d_2 are the respective degradation rates. p_1 to p_{12} are the rate constants. Even at the micro-level of the elements shown in Figure 1.2, optimal performance requires a balance between numerous processes governed by coupled equations of the form of equation [1.1].

System biology modeling approaches address such complex interactions between components of the mitochondria, leading to the fission–fusion cycles, as well as oscillations in concentrations of ATP and Ca^{2+}. They also address how small perturbations in biochemical concentrations result in very different fates, implying that bistability exists. Such bistability indicates "choices" in the biological processes. Feedback and feedforward loops exist to control and optimally balance the fission–fusion cycles, as well as the ATP production machinery.

Furthermore, the key proteins involved in mitochondrial dynamics are regulated in direct response to the bioenergetic state of the mitochondria. Three members of the dynamin family, which are GTPase enzymes, critical for many cell functions, are key components of the fission and fusion machineries (van der Bliek *et al.* 2013). We still have only a limited knowledge of the mitochondrial proteome, its entire set of proteins, but expect it to be customized and optimized to the location in the body.

Mitochondrial health (Patel *et al.* 2013) has been linked to its membrane potential, making it useful as an equivalent measure of health, an indicator of the effectiveness of the electron transport chain, the number of ATPases using the membrane potential, the proton leaks across the inner mitochondrial membrane and other geometric factors. Optimal mitochondrial health is linked to autophagy and fusion, the balance between fission and fusion, and the density of mitochondria in the cell.

Existing limited models of abnormal mitochondrial dynamics are insufficient to explain phenotypic variability – the aggregate of an organism's observable characteristics – in symptoms. The mechanisms of mitochondrial functions across multiple levels of organization – molecular and organelle levels – are needed (Eisner 2018). The current, mostly descriptive representations cannot accurately model multivariate dynamics since physiological and pathological processes result from biochemical, morphological and mechanical dynamics at more than one scale, and we do not fully understand these.

The primary sites of neuronal energy consumption are at the synapses (pre and post), where mitochondria need to congregate and adapt to local energy needs. They do this via feedforward and feedback regulatory mechanisms known as mitochondrial plasticity, where adaptations to neuronal energy states occur via changes in morphology, function and position (Rossi and Pekkurnaz 2019). Mitochondrial distribution and dynamics are regulated at the molecular level by mitochondrial and axonal cytoskeleton tracks. A motor–adaptor complex exists on the mitochondrial surface that contains the transport motor proteins kinesin and dynein. Proteins Miro and Milton mostly govern mitochondrial movement (Schwarz 2013), and multiple signaling pathways converge to tailor mitochondrial positioning (Rossi and Pekkurnaz 2019). A constrained optimization framework may be used to model mitochondrial movement via an evolutionarily refined weighting mechanism. Similarly, immediate energy needs at synapses result in mitochondrial plasticity, in a mechanism for the constrained optimization of energy availability and use, where it is most needed.

The matching of energy supply to demand is evolutionarily conserved, where the organelle and the cell must optimize energy use locally and globally to ensure that balance. Where there are shortfalls in energy, the "optimal" choice may become a dysfunctional pathway, resulting in pathologies and neurodegenerative diseases. Mathematical models that incorporate data can represent the optimal choices and be powerful tools for a systematic understanding.

1.4. Mechanical aspects

The mechanical aspects of cellular, mitochondrial and organelle dynamics are coupled with numerous biochemical processes that are related to healthy cell functioning, as well as to pathological developments (Feng and Kornmann 2018). Examples include mechanosensitive ion channels, curvature sensing proteins and force sensing by the cytoskeleton and plasma membrane. Organelles exchange metabolites and information by moving around the cytoplasm and coming into physical contact with each other. This intermingling appears to be an important feature for the proper functioning of eukaryotic cells. Biological phenomena observed at organelle contact locations, for example, reticulum-induced mitochondrial fission, are at least partially attributable to mechanical stimulation (Feng and Kornmann 2018).

In addition, many chemicals can alter the mechanical properties of living cells (Lim *et al.* 2006), indicating that certain cellular mechanical properties can be used as indicators of health. An understanding of mechanosensitive signaling pathways is fundamental (Moeendarbary and Harris 2014; Petridou *et al.* 2017) for the development of clinical diagnostics, as well as therapeutically successful interventions.

The mitochondria connect physically with other organelles within the cell. These interactions occur randomly in part, but are also driven by the microenvironment. The endoplasmic reticulum (ER) (also a tubular organelle) and the mitochondria tether to each other via interacting proteins situated on opposing membrane faces. Reciprocal communications transmit danger signals that can trigger multiple, synergistic responses. If needed, the number of ER–mitochondrial contact sites can be increased to allow for enhanced molecular transfers. This interface provides a platform for the regulation of different processes, such as the coordination of calcium transfers, the regulation of mitochondrial dynamics, the regulation of inflammasome formation, morphological changes and the provision of membranes for autophagy (Giorgi *et al.* 2009; Marchi *et al.* 2014).

One example of such critical coupling is the oscillations between Ca^{2+} concentrations in the mitochondria and the ER (Figure 1.3). These concentrations are a ubiquitous intracellular signaling mechanism for numerous cell functions. As examples, we cite neurotransmitter release from neurons and astrocytes, and metabolic processes. Interestingly, signaling information is stored in the oscillation characteristics, in particular, frequency, amplitude and duration. The aforementioned coupling, i.e. the crosstalk of Ca^{2+} ions, occurs within an optimal microdomain, with approximately 50 nm of spacing between a receptor and a uniporter – a membrane protein that is specialized to transport a particular substrate species across a cell membrane. At a critical distance, an optimal amount of Ca^{2+} released by the ER is taken up by the mitochondria, resulting in the successful generation of Ca^{2+} signals in healthy cells (Qi *et al.* 2015, see Figures 1.4 and 1.5).

In this study, optimal microdomain distances were found by varying parameter values in first-order Ca^{2+} flux differential equations (represented in the concentration flows in Figure 1.3(c)). The dynamic behavior is a result of a constrained optimization.

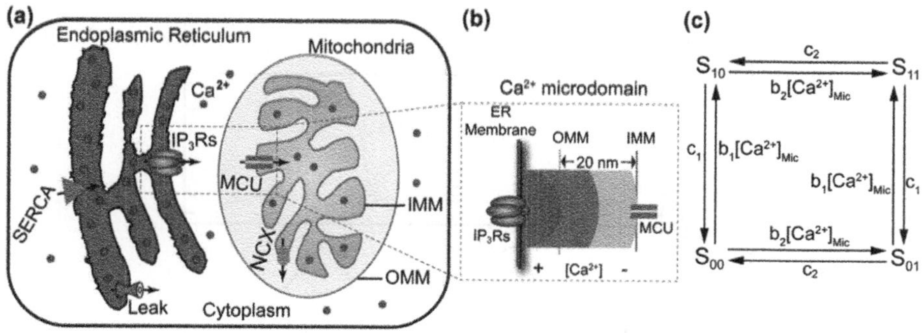

Figure 1.3. *The schematic diagram of the components and fluxes included in the crosstalk model between endoplasmic reticulum and mitochondria for Ca^{2+} oscillations: (a) endoplasmic reticulum (ER) and mitochondria in the cytoplasm, (b) microdomain showing activity between the ER and mitochondria, (c) four-state model with binding and unbinding rates (Qi et al. (2015), with permission). For a color version of this figure, see www.iste.co.uk/challamel/mechanics3.zip*

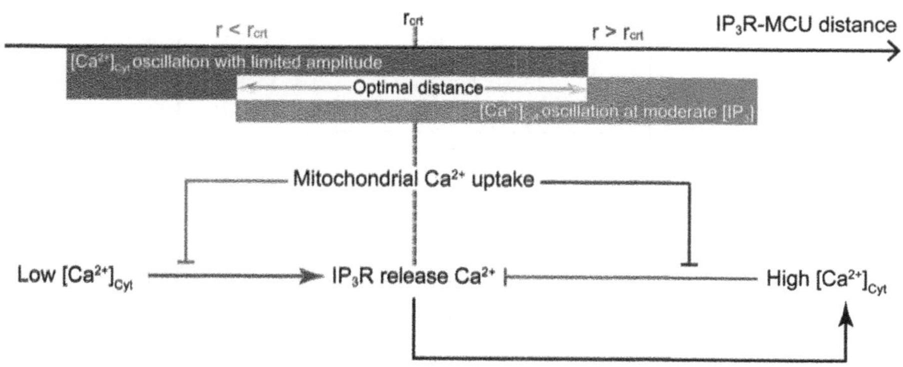

Figure 1.4. *Schematic representation of how mitochondria modulate $[Ca^{2+}]_{Cyt}$. Identifying the critical distance at which 50% of the IP_3R-released Ca^{2+} ions are taken up by mitochondria. Outside this range, negative and positive control occurs on the Ca^{2+} (Qi et al. (2015), with permission). For a color version of this figure, see www.iste.co.uk/challamel/mechanics3.zip*

Figure 1.4 schematically represents an optimal microdomain distance for Ca^{2+} uptake. Figure 1.5(a) and (b) shows the Ca^{2+} fluctuations for cases with (a) and without (b) the presence of mitochondria.

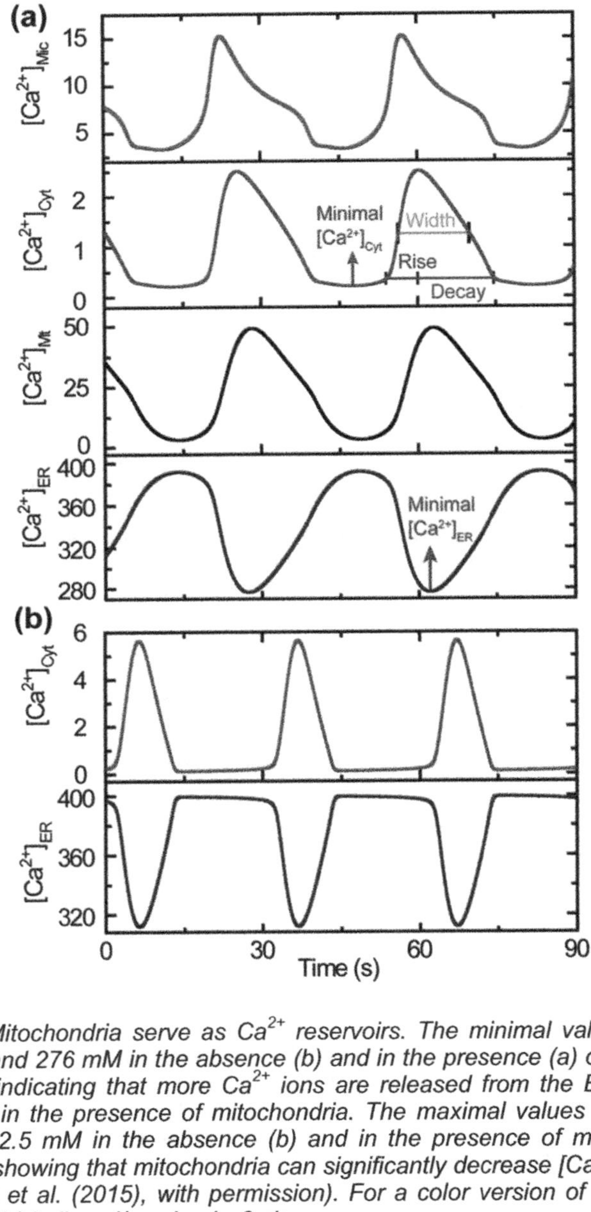

Figure 1.5. *Mitochondria serve as Ca^{2+} reservoirs. The minimal values of $[Ca^{2+}]_{ER}$ are 311 mM and 276 mM in the absence (b) and in the presence (a) of mitochondria, respectively, indicating that more Ca^{2+} ions are released from the ER during each spiking cycle in the presence of mitochondria. The maximal values of $[Ca^{2+}]_{Cyt}$ are 5.6 mM and 2.5 mM in the absence (b) and in the presence of mitochondria (a), respectively, showing that mitochondria can significantly decrease $[Ca^{2+}]_{Cyt}$ oscillation amplitude (Qi et al. (2015), with permission). For a color version of this figure, see www.iste.co.uk/challamel/mechanics3.zip*

Mitochondria also have bidirectional communications with other cellular organelles, in particular, lysosomes (spherical vesicles that contain enzymes that can break down many kinds of biomolecules) and peroxisomes (organelles involved in the catabolism of various acids, in addition to other tasks). These communications are involved in the pathology of mitochondrial diseases (Diogo *et al.* 2018). These communications also provide coupling between the organelles and, while having evolved for efficiencies and survival, can also cause coupled dysfunctions. For example, lysosomes and peroxisomes are affected structurally and functionally by genetic defects in mitochondrial proteins that are known as mitochondrial diseases. Similarly, lysosomal and peroxisomal diseases perturb mitochondria. Lysosomal storage diseases perturb peroxisomal metabolism and mitochondrial function. Peroxisomal diseases often lead to alterations in mitochondrial structure, redox (ROS) balance and metabolism. The saturation of lysosomal capacity is often observed in mitochondrial diseases, with an accumulation of dysfunctional lysosomes and autophagosomes. Such couplings challenge both the modeling and the development of focused clinical treatments. Given all these couplings, we hypothesize that there can be a variety of optimal behavioral regimes that avoid pathological pathways. A goal can be to identify these via dynamic governing models.

In summary, an appropriate set of optimal interaction mechanisms need to be in place for healthy cell-wide functioning. In this context, a full understanding of cellular metabolism and mitochondrial diseases requires an understanding of mitochondrial communications with the rest of the cell, as well as within the organelle.

1.5. Mitochondrial motility

Mitochondrial motility, or trafficking, is critical for the survival of all cells. Neurons, which extend their axons and dendrites up to a meter, between two and three orders of magnitude greater than most other cells, are especially vulnerable to any inability by their mitochondria to get to sites with a high energy demand. At any instant of time, about 10–40% of the mitochondria are generally moving, with about half of those moving away from the cell body (anterograde, kinesin-dependent) and the rest towards the cell body (retrograde, dynein-dependent) (Schwarz 2013).

The distribution of mitochondria over long distances in the neurons is regulated by a complex molecular machinery that has evolved to match the very dynamic demand for energy with an optimal mitochondrial distribution (Schwarz 2013). Highly branched paths of the complex neuron geometry must be navigated, knowing where and when to stop. Machinery for fission and fusion can intersect with machinery for motility using feedforward and feedback mechanisms. Misregulation of motility can lead to neurodegeneration (Vanhauwaert *et al.* 2019).

The dendritic synapses are where the neurons receive signals from other neurons. Energy demand is greatest at the synapses and can change rapidly in response to almost instantaneous environmental changes. The same is true at the axonal synapses, which are used by neurons to transfer signaling downstream to subsequent neurons. Neuronal axons lie flat and are typically about a micron in diameter. Trafficking is along linear arrays of uniformly polarized microtubules, where the negative ends of the tubules are anchored in the cell body and the positive charge ends in the distal tips. Due to the local morphology, axonal mitochondria have separated from the reticulum and exist as discrete organelles of a dimension, typically 1–3 microns, with those in dendrites tending to be longer.

Mitochondria are a fundamental component for healthy living, supporting optimal functioning and efficient energy usage at all levels, thus avoiding numerous pathologies. In the discussion so far, we maintain that a significant controlling aspect of mitochondrial functioning is based on optimizations of a variety of defining characteristics. Here, motility and the placement of mitochondria within dendrites and axons can be viewed as optimal solutions, assuring sufficient energy at locations of high energy demand.

1.6. Cristae, ultrastructure and supercomplexes

Mitochondrial dynamics has evolved to define a broad array of actions and activities: already discussed are fission and fusion, and also cristae modifications, the constant changing of shape at the macro- and ultrastructural levels. Such modifications in shape are directly connected to their bioenergetic function. The cristae, where the respiratory chain complexes exist, have junctions of width between 20 nm and 40 nm. They control metabolite and protein access to the interior volume, known as the cristae lumen. They also control the apoptosis process of cell death. The junction structure and lumen are modified for more efficient respiration. So-called supercomplexes are created as part of the cristae remodeling, and these are required for optimal mitochondrial respiration. Slight morphological changes can fine-tune the efficiency of energy-producing respiration. There is a correlation between mitochondrial respiratory capacity and the number of cristae (Baker *et al.* 2019). A clear understanding of the supercomplexes, as well as the basis upon which cristae remodeling is optimized, can provide us with clues for clinical interventions that target respiratory-based diseases. We can hypothesize that morphological, mechanical and biochemical aspects play a role in cristae remodeling optimization, with the following discussion being of relevance.

The inner mitochondrial membrane and the cristae structure have only recently started to be understood. The inner membrane has the larger area even though it is enclosed within an outer membrane of a smaller area. To accommodate this, broad folds, called cristae, have evolved and project into the mitochondrial matrix where the bioenergetic processes occur. The observed morphology of the inner mitochondrial membrane is the result of the minimization of the system's free energy. This free energy is a function of local bending energy and curvatures, the surface area, pressure differences and surface tension exerted by proteins. It appears, based on analytical models of these characterizing parameters, that the surface tension and tensile forces act as constraints on the morphology. The varied shapes of the cristae are defined by the stationary states of the above minimization. It appears that motor protein tensile forces are responsible for a stable inner membrane shape (Ghochani *et al.* 2010). The nanoscale cristae, regulated by molecular mechanisms, influence mitochondrial function (Mannella *et al.* 2013).

The tight coupling between cellular bioenergetics, metabolism, the inner membrane structure and mitochondrial function, as well as a continuous and high level of neuronal energy requirements, suggests links to understanding neurodegenerative diseases, bioenergetic dysfunction and mitochondrial diseases, and hopefully, to the identification and the enaction of clinical efforts for recovery. Such complex coupling suggests that constrained optimal decisions are a significant aspect of the governing of system behavior.

1.7. Mitochondrial diseases and neurodegenerative disorders

Mitochondrial dysfunction and disease are fundamental to the progression of the major neurodegenerative and neuropsychiatric diseases. Causes can include genetic defects, and intra- and extracellular environmental instigators, resulting in incurable neurodegeneration with motor, behavioral and cognitive losses of functioning, leading to death (Correia and Moreira 2018).

Primary mitochondrial defects affect all aspects of functioning. They are linked to diseases such as Alzheimer's, Huntington's, cancer, in the aging process (Lemonde and Rahman 2014) and are involved in the pathogenesis of multiple sclerosis (Adiele and Adiele 2019). Even relatively minor mitochondrial dysfunction can lead to Parkinson's disease and Huntington's disease, which also has psychiatric manifestations (Buhlman 2016).

Mitochondrial energy production occurs within the inner membrane, in the respiratory chain within the five enzymatic complexes that are housed in the matrix (see Figure 1.1). Nuclear DNA (nDNA) and the mitochondrial DNA (mtDNA) control the respiratory chain. Each cell has hundreds or thousands of mtDNA copies, and during cell division, any genetic mutations in mtDNA are distributed to the daughter cells randomly, resulting in both mutant- and wild (non-mutant)-type mtDNA copies. With further cell divisions, there is an accumulation of mutant cells.

When the ratio of mutant to wild-type mtDNA copies grows beyond a certain threshold, which is random within a range, clinical symptoms of diseases can occur due to a misfiring of energy production. This group of diseases is known as mitochondrial disease. Different patients manifest their diseases at different stages, in different ways, due to the randomness of the relation between the above ratio and how the organism responds (Kurt and Topal 2013).

By way of fusion, damaged mtDNA is diluted, lowering the ratio between mutant and wild types below the threshold. Fission provides a mechanism to isolate components that become damaged due to age, or due to increased oxidative stress, for elimination. An imbalance between fission and fusion, discussed earlier as an optimal balancing, results in mitochondrial dysfunction (Panchal and Tiwari 2019).

Aging can lead to an accumulation of ROS, major contributors to oxidative stress, due to the imbalance between the production of ROS and their oxidation, which then affects mitochondrial respiratory chain function. Cumulative oxidative stress is viewed as the critical factor that precedes a cascade to dysfunction (Elfawy and Das 2019).

Mitochondrial diseases have a major relevance to military personnel health. For example, approximately one-third of the 1990–1991 US Gulf War veterans – in the range of 175,000–250,000 soldiers – developed chronic multisymptom health problems, known by the term "Gulf War Illness", with mechanisms that adversely affect mitochondria (Koslik *et al.* 2014; Chen *et al.* 2017).

1.8. Modeling

More generally, and rounding out the above discussion, when compared to the tremendous number of papers that describe experimental results and hypothesize causes, effects and couplings, there have been a comparatively small number of works on the mathematical modeling of various aspects of mitochondrial function. Early examples of such models (Magnus and Keizer 1997, 1998a, 1998b) are quite complex, with simpler derivative models attempted (Bertram *et al.* 2006; Saa and Siqueira 2013). Derived equations govern ATP production during glucose metabolism via oxidative phosphorylation. There have also been efforts to relate mitochondrial dynamics to neuronal spike generation, providing a powerful tool for disease modeling and potentially enabling clinical interventions (Venkateswaran *et al.* 2012). A computational model for the mitochondrial respiratory chain is derived to appropriately balance mass, charge and free energy transduction (Beard 2005). Evidence exists of optimal decisions occurring within the mitochondria. The challenge of formulating even a simple model is a serious undertaking.

Mathematical modeling has been attempted for other critically related aspects of energetics, but are beyond our scope here, which has been to provide background on the mitochondria, where optimizations appear to have been the governing framework for processes.

Due to the formidable challenges in gathering biological data, understanding cause and effect, and even identifying all the components and constituents of cells, it is fully understandable and logical that these have taken precedence in the research community. As these aspects have become better understood and deeper, we have seen that mathematical modeling efforts have started to take hold and bear fruit.

1.9. Concluding summary

Mitochondria play a central role in cellular energetics, and are involved in almost all neurological disorders. Thus, mitochondrial function is at the core of our understanding of neurological health. How dysfunction leads to many of the most debilitating and fatal neurodegenerative diseases is still poorly understood. Some understanding may be garnered if the processes that occur within the cell can be placed in an optimization framework, which may then lead to therapeutic interventions to improve the health of the many millions of people who suffer from these diseases.

Mitochondria move, stop and anchor, undergo fusion and fission, and degrade on the spatial and temporal scales needed to meet the broad spectrum of cellular energy needs, as well as Ca^{++} buffering needs, throughout the neuron. Any lack of these functions can lead to mitochondrial and neuronal death. Based on the literature, of which only a very small fraction is referenced here, it appears that there are energy-based mechanisms that operate locally (sub-optimizations) and globally across multiple cells and beyond (perhaps full optimizations). Such a constrained optimization framework appears to be a reasonable way to better understand very complex subcellular processes.

It appears that among numerous progression paths, some lead to dysfunction, in particular, due to ineffective energy production. These paths, we believe, satisfy some optimalities in order to become the chosen progression path. We may conclude that physiological, as well as pathological paths, in conjunction with other environmental factors, can be optimal choices for the organism. Perhaps, depending on the combination of intracellular and extracellular environments, such optimizations can be based on different cellular/subcellular states, characteristics and constraints. Clearly, due to the criticality of energy production and usage, we expect that these are fundamental aspects of any optimal decision-making. There may not be any global optimizations, only local sub-optimizations. Understanding these optimality decisions may provide clues and eventually paths for clinical interventions and cures for some of humanity's most serious neurodegenerative diseases.

Approaching such problems with the view of an applied mechanician may provide new perspectives on cell modeling and intracellular processes. We trust that this chapter will motivate others to try their hand at biological modeling.

1.10. Acknowledgments

It is a pleasure to thank the editors of this volume for their effort in organizing the authors and chapters, and for considering a chapter on a topic that may be considered outside the mainstream focus of this book, as well as the very useful feedback on the first draft of this chapter.

1.11. Appendix

This volume acknowledges the professional life and accomplishments of Isaac Elishakoff. I knew of his work before I met him in 1982, when I gave a talk at the Technion, in Haifa, where Isaac was a faculty member. He was very kind to a novice; at the time I worked for a consulting engineering firm in New York. It would be another seven years before I would join Rutgers University. But during that consulting period, I continued my friendship with Isaac, and we and two additional colleagues published a paper on the use of MACSYMA, one of the early symbolic codes, in problems of random vibration.

Isaac and I continued our near-periodic interactions, randomly exchanging ideas and holiday greetings. He eventually came to Florida Atlantic University, not missing a beat in his extraordinary productivity. He writes books as others write papers. His interest in mechanics goes beyond the technical aspects, intersecting with the historical provenance of the fundamental ideas that shape our disciplines. His works continue to add insights and dimension to our understanding of complex physical processes.

While my professional interests very much overlap with Isaac's, I have recently become interested in the biological sciences, in particular, brain energetics, and whether we can apply some of our engineering and mathematical modeling skills to the fantastic and very complex processes, by which the cells in our bodies create energy. I am pleased to honor Isaac by presenting a summary of some interesting aspects of the functioning of the mitochondria, an organelle that exists in large numbers in most of our cells. It creates the energy that our bodies require in order to live, survive and think. This community has much to offer in helping to increase our understanding of these beyond-complicated processes. I am sure that Isaac would agree.

Congratulations Isaac, for what you have achieved so far, and for what you will continue to contribute!

1.12. References

Adiele, R.C. and Adiele, C.A. (2019). Metabolic defects in multiple sclerosis. *Mitochondrion*, 44, 7–14.

Baker, N., Patel, J., Khacho, M. (2019). Linking mitochondrial dynamics, cristae remodeling and supercomplex formation: How mitochondrial structure can regulate bioenergetics. *Mitochondrion*, 49, 259–268.

Beard, D.A. (2005). A biophysical model of the mitochondrial respiratory system and oxidative phosphorylation. *PLoS Comput. Biol.*, 1(4), 252–264.

Benard, G., Bellance, N., James, D., Parrone, P., Fernandez, H., Letellier, T., Rossignol, R. (2007). Mitochondrial bioenergetics and structural network organization. *J. Cell Sci.*, 120(5), 838–848.

Benaroya, H. (2020). Brain energetics, mitochondria, and traumatic brain injury. *Rev. Neurosci.* [Online]. Available at: https://doi.org/10.1515/revneuro-2019-0086.

Bertram, R., Pedersen, M.G., Luciani, D.S., Sherman, A. (2006). A simplified model for mitochondrial ATP production. *J. Theor. Biol.*, 243, 575–586.

Buhlman, L.M. (2016). *Mitochondrial Mechanisms of Degeneration and Repair in Parkinson's Disease*. Springer Nature, Cham, Switzerland.

Castora, F.J. (2019). Mitochondrial function and abnormalities implicated in the pathogenesis of ASD. *Prog. Neuropsychopharmacol. Biol. Psychiatry*, 92, 83–108.

Chan, F., Lax, N.Z., Voss, C.M., Aldana, B.I., Whyte, S., Jenkins, A., Nicholson, C., Nichols, S., Tilley, E., Powell, Z., Waagepetersen, H.S., Davies, C.H., Turnbull, D.M., Cunningham, M.O. (2019). The role of astrocytes in seizure generation: Insights from a novel *in vitro* seizure model based on mitochondrial dysfunction. *Brain*, 142, 391–411.

Chauhan, A., Vera, J., Wolkenhauer, O. (2014). The systems biology of mitochondrial fission and fusion and implications for disease and aging. *Biogerontology*, 15, 1–12.

Chen, Y., Meyer, J.N., Hill, H.Z., Lange, G., Condon, M.R., Klein, J.C., Ndirangul, D., Falvo, M.J. (2017). Role of mitochondrial DNA damage and dysfunction on veterans with Gulf War Illness. *PLoS One*, 12(9), e0184832.

Correia, S.C. and Moreira, P.I. (2018). Role of mitochondria in neurodegenerative diseases: The dark side of the "energy factory". In *Mitochondrial Biology and Experimental Therapeutics*, Oliviera, P.J. (ed.). Springer Nature, Cham, Switzerland.

Diogo, C.V., Yambire, K.F., Mosquera, L.F., Branco, T., Raimundo, N. (2018). Mitochondrial adventures at the organelle society. *Biochem. Biophys. Res. Commun.*, 500, 87–93.

Eisner, V., Picard, M., Hajnóczky, G. (2018). Mitochondrial dynamics in adaptive and maladaptive cellular stress responses. *Nat. Cell Biol.*, 20, 755–765.

Elfawy, H.A. and Das, B. (2019). Crosstalk between mitochondrial dysfunction, oxidative stress, and age related neurodegenerative disease: Etiologies and therapeutic strategies. *Life Sci.*, 218, 165–184.

Elishakoff, I. and Zingales, M. (2003). Contrasting probabilistic and anti-optimization approaches in an applied mechanics problem. *Int. J. Solids Struct.*, 40, 4281–4297.

Elishakoff, I., Haftka, R.T., Fang, J. (1994). Structural design under bounded uncertainty – Optimization with anti-optimization. *Comp. Struct.*, 53(6), 1401–1405.

Feng, Q. and Kornmann, B. (2018). Mechanical forces on cellular organelles. *J. Cell Sci.*, 131, 1–9.

Ghochani, M., Nulton, J.D., Salamon, P., Frey, T.G., Rabinovitch, A., Baljon, A.R.C. (2010). Tensile forces and shape entropy explain observed crista structure in mitochondria. *Biophys. J.*, 99, 3244–3254.

Giorgi, C., De Stefani, D., Bononi, A., Rizzuto, R., Pinton, P. (2009). Structural and functional link between the mitochondrial networks and the endoplasmic reticulum. *Int. J. Biochem. Cell Biol.*, 41(10), 1817–1827.

Kembro, J.M., Aon, M.A., Winslow, R.L., O'Rourke, B., Cortassa, S. (2013). Integrating mitochondrial energetics, Redox and ROS metabolic networks: A two-compartment model. *Biophys. J.*, 104, 332–343.

Koslik, H.J., Hamilton, G., Golomb, B.A. (2014). Mitochondrial dysfunction in Gulf War Illness revealed by 31Phosphorus magnetic resonance spectroscopy: A case-control study. *PLoS One*, 9(3), e92887.

Kurt, B. and Topal, T. (2013). Mitochondrial disease. *Dis. Mol. Med.*, 1(1), 11–14.

Lackner, L.L. (2014). Shaping the dynamic mitochondrial network. *BMC Biol.*, 12, 35.

Lemonde, H. and Rahman, S. (2014). Inherited mitochondrial disease. *Pediatr. Child Health*, 25(3), 133–138.

Lim, C.T., Zhou, E.H., Quek, S.T. (2006). Mechanical models for living cells – A review. *J. Biomech.*, 39, 195–216.

Mannella, C.A., Lederer, W.J., Jafri, M.S. (2013). The connection between inner membrane topology and mitochondrial function. *J. Mol. Cell Cardiol.*, 62, 51–57.

Marchi, S., Patergnani, S., Pinton, P. (2014). The endoplasmic reticulum–mitochondria connection: One touch multiple functions. *Biochim. Biophys. Acta*, 1837, 461–469.

Moeendarbary, E. and Harris, A.R. (2014). Cell mechanics: Principles, practices, and prospects. *WIRE's Sys. Bio. Med.*, 6, 371–388.

Pagliuso, A., Cossart, P., Stavru, F. (2018). The ever-growing complexity of the mitochondrial fission machinery. *Cell Mol. Life Sci.*, 75, 355–374.

Panchal, K. and Tiwari, A.K. (2019). Mitochondrial dynamics, a key executioner in neurodegenerative diseases. *Mitochondrion*, 47, 151–173.

Patel, P.K., Shirihai, O., Huang, K.C. (2013). Optimal dynamics for quality control in spatially distributed mitochondrial networks. *PLOS Comput. Biol.*, 9(e1003108), 1–11.

Petridou, N.I., Spiró, Z., Heisenberg, C.-P. (2017). Multiscale force sensing in development. *Nat. Cell Biol.*, 19(6), 581–588.

Qi, H., Li, L., Shuai, J. (2015). Optimal microdomain crosstalk between endoplasmic reticulum and mitochondria for Ca^{2+} oscillations. *Scientific Reports*, 5(7984). DOI:10.1038/srep07984.

Qiu, Z. and Elishakoff, I. (2001). Anti-optimization technique – A generalization of interval analysis for nonprobabilistic treatment of uncertainty. *Chaos Soliton Fract.*, 12, 1747–1759.

Rossi, M.J. and Pekkurnaz, G. (2019). Powerhouse of the mind: Mitochondrial plasticity at the synapse. *Curr. Opin. Neurobiol.*, 57, 149–155.

Saa, A. and Siqueira, K.M. (2013). Modeling ATP production in mitochondria. *Bull. Math. Biol.*, 75, 1636–1651.

Schwarz, T.L. (2013). Mitochondrial trafficking in neurons. *CSH Perspect. Biol.*, 5, a011304.

Simcox, E.M. and Reeve, A.K. (2016). An introduction to mitochondria, their structure and functions. In *Mitochondrial Dysfunction in Neurodegenerative Disorders*, Reeve, A.K., Simcox, E.M., Duchen, M.R., Turnbull, D.M. (eds). Springer Nature, Cham, Switzerland.

Skulachev, V.P. (2001). Mitochondrial filaments and clusters as intracellular power-transmitting cables. *Trends Biochem. Sci.*, 26(1), 23–29.

Tzameli, I. (2012). The evolving role of mitochondria in metabolism. *Trends Endocrinol. Metab.*, 23(9), 417–419.

Vakifahmetoglu-Norberg, H., Ouchida, A.T., Norberg, E. (2017). The role of mitochondria in metabolism and cell death. *Biochem. Bioph. Res. Co.*, 482, 426–431.

Van der Bliek, A.M., Shen, Q., Kawajiri, S. (2013). Mechanisms of mitochondrial fission and fusion. *CSH Perspect. Biol.*, 5, a011072.

Vanhauwaert, R., Bharat, V., Wang, X. (2019). Surveillance and transport of mitochondria in neurons. *Curr. Opin. Neurobiol.*, 57, 87–93.

Venkateswaran, N., Sekhar, S., Sanjayasarathy, T.T., Krishnan, S.N., Kabaleeswaran, D.K., Ramanathan, S., Narayanasamy, N., Jagathrakshakan, S.S., Vignesh, S.R. (2012). Energetics based spike generation of a single neuron: Simulation results and analysis. *Front. Neuroenerg.*, 4(2), 1–12.

Vinogradskaya, I.S., Kuznetsova, T.G., Suprunenko, E.A. (2014). Mitochondrial network of skeletal muscle fiber. *Mosc. U. Biol. Bull.*, 69(2), 57–66.

Wai, T. and Langer, T. (2016). Mitochondrial dynamics and metabolic regulation. *Trends Endocrinal. Metab.*, 27(2), 105–117.

Westermann, B. (2012). Bioenergetic role of mitochondrial fusion and fission. *Biochim. Biophys. Acta.*, 1817, 1833–1838.

Zick, M. and Reichert, A.S. (2011). *Mitochondria, in Cellular Domains*, Nabi, I.R. (ed.). John Wiley & Sons, Chichester, England, 87–111.

2

The Concept of Local and Non-Local Randomness for Some Mechanical Problems

2.1. Introduction

In many mechanical problems, the uncertainties in defining geometrical and/or material properties lead to modeling the latter as random fields (RFs). RFs are characterized, besides the single-point statistics, as the mean value and the variance, as well as by the correlation functions that give the level of stochastic correlation of the RFs evaluated at two, or more, points. Even for equal single-point statistics, the response of any mechanical problem may strongly depend on the kind of correlation functions defining the RFs (Vanmarcke 2010).

One way to classify the different classes of correlation functions available in the literature is based on the ability to capture the fractal and Hurst effects. Generally, the fractal dimension, D, is described as a roughness parameter, i.e. a local behavior of the RF; while the Hurst parameter, H, describes the long-range persistence, i.e. a non-local (global) behavior of the RF. Many papers dealing with the analysis of fractal and Hurst effects in some mechanical problems, such as in fracture surfaces (Turcotte 1997; Laudani and Ostoja-Starzewski 2020), turbulence flows (Scotti *et al.* 1995; Jaw and Chen 1999; Laudani *et al.* 2020) and random vibrations (Shen *et al.* 2015), can be found in the literature in the last decades.

Supposing the flexural deformability of the beams to be a Gaussian homogeneous RF, the investigation of the concept of spatial randomness in local and non-local mechanical problems is the aim of this work. Although these kinds of investigations

---

Chapter written by Giovanni FALSONE and Rossella LAUDANI.

are not new in the literature (Demmie and Ostoja-Starzewski 2016), here, a mathematical similarity between them and the non-local integral constitutive equations (Eringen 1983) will be shown evidencing analogy between the kernel function considered in the constitutive equations and the correlation function used for representing the statistic dependence level between the deformabilities measured at two different sections of beams. Observations about the properties of the kernel function in the non-local constitutive equations are also not new (Ghavanloo *et al.* 2019), i.e. the spatial non-locality deals with long-range interaction and the kernel function should converge to the Dirac delta function in order to reduce the non-local elasticity to the classical (local) elasticity.

Here, a link between the statistical RF theory and the local and non-local randomness in stochastic mechanics is introduced. In fact, in the literature, it is possible to find some kinds of correlation functions that allow the capture of the local behavior of the RF (White Noise, Matérn and Powered exponential correlation function) or both their local and non-local characteristics (Cauchy correlation function). The use of these classes of correlation functions has made it possible to investigate, from a probabilistic point of view, how much the local and non-local characteristics of the flexural deformability affect the level of randomness of the various stochastic response quantities, such as the displacement and/or the internal forces of the indeterminate beams.

The sensitivity of the stochastic response quantities to the local and non-local randomness dependence of the flexural deformability will be investigated through some examples of statically determinate and indeterminate stochastic beams, under different conditions of load and constraint. In particular, the response stochastic quantities will be found thanks to the use of the probability transformation method (PTM) (Falsone and Settineri 2013a, 2013b; Falsone and Laudani 2018, 2019a, 2019b). It is important to note that, for statically determinate beams, the cinematic responses (transversal displacements and rotations) are random. In comparison, for indeterminate beams, the internal force responses (shear and bending moment) are also random (Elishakoff *et al.* 1995, 1999).

2.2. Preliminary concepts

With the aim to examine the dependence of the stochastic response on the kind of correlation functions defining the material/geometric RFs, in the following sections, some closed-form solutions of statically and redundantly stochastic bending beams are described.

2.2.1. *Statically determinate stochastic beams*

The beam under consideration has a linear axis, is constrained in such a way that it is statically determinate and is characterized by a flexural deformability (i.e. the inverse of the flexural stiffness) that is supposed to be a Gaussian homogeneous RF given by:

$$\Phi(z) = \frac{1}{EI(z)} = \Phi_0(1 + \alpha(z)) \qquad [2.1]$$

where Φ_0 is the mean value of $\Phi(z)$, while the correlation is represented as:

$$\sigma_\Phi^2(z_1, z_2) = \sigma_\Phi^2(|z_1 - z_2|) = \Phi_0^2\sigma_\alpha^2(|z_1 - z_2|) \qquad [2.2]$$

where the dependence of the correlation function on the distance between z_1 and z_2 is related to the assumption of the homogeneous random field. However, from equation [2.1], it is important to emphasize that, if the uncertainty is defined through the stiffness, we can always apply a simple PTM for obtaining the probabilistic characterization of the deformability. For any external load condition, from the equilibrium conditions, it is always possible to find the bending moment function $M(z)$, that is deterministic, if the load and the constraints are deterministic. This is due to the fact that $M(z)$ does not depend on the beam deformability.

The value of the transversal displacement $u(\bar{z})$, at any abscissa $z = \bar{z}$, can be obtained by the application of the principle of virtual work in the form:

$$u(\bar{z}) = \int_0^l \Phi(z)M(z)M^{(1)}(z, \bar{z})dz \qquad [2.3]$$

where $M^{(1)}(z, \bar{z})$ is the bending moment law in the beam due to a unitary transversal load acting at $z = \bar{z}$. As well as $M(z)$, the function $M^{(1)}(z, \bar{z})$ is deterministic. Consequently, [2.1]–[2.3] show that the transversal displacement $u(\bar{z})$ is a Gaussian random variable characterized by the following mean value and variance:

$$\mu_u(\bar{z}) = \Phi_0 \int_0^l \Phi(z)M(z)M^{(1)}(z, \bar{z})dz, \qquad [2.4]$$

$$\sigma_u^2(\bar{z}) = \Phi_0^2 \times$$

$$\int_0^l \int_0^l M(z_1)M(z_2)M^{(1)}(z_1, \bar{z})M^{(1)}(z_2, \bar{z})\sigma_\alpha^2(|z_1 - z_2|)dz \qquad [2.5]$$

Depending on the type of law of $\sigma_\alpha^2(|z_1 - z_2|)$, and of the laws of $M(z)$ and $M^{(1)}(z, \bar{z})$, the previous integrals can be easily solved and, often, a closed-form solution can be obtained. Hence, for statically determinate stochastic beams, the probabilistic characterization of displacements does not show any particular difficulty. This result was evidenced in some of Elishakoff's works since 1995 (Elishakoff *et al.* 1995, 1999).

2.2.2. *Statically indeterminate stochastic beams*

In this section, it will be shown that it is possible to obtain a closed-form solution, for some statically indeterminate stochastic beams, too. To explain the used approach, the simple stochastic beam represented in Figure 2.1(a) is taken into consideration. It is characterized by only one redundant force and, with the aim of applying the force method for solving the indetermination, the scheme represented in Figure 2.1(b) is considered. The redundant force X is obtained by imposing the cinematic constraint condition $u_A = 0$. In particular, it is possible to write:

$$u_A = u_A^{(q)} + u_A^{(X)} = 0 \qquad [2.6]$$

with $u_A^{(q)}$ and $u_A^{(X)}$ being the contribution to the displacement due to the external load and redundant force, respectively, in the scheme of Figure 2.1(b), i.e.:

$$u_A^{(q)} = \frac{q}{2} \int_0^l z^3 \Phi(z) dz = \frac{q}{2} A, \qquad [2.7]$$

$$u_A^{(X)} = X \int_0^l z^2 \Phi(z) dz = XB \qquad [2.8]$$

where A and B are two Gaussian random variables defined by the following mean values and variances:

$$\mu_A = \frac{\Phi_0 l^4}{4}; \quad \sigma_A^2 = \Phi_0^2 \int_0^l \int_0^l z_1^3 z_2^3 \sigma_\alpha^2(|z_1 - z_2|) dz_1 dz_2 \qquad [2.9]$$

$$\mu_B = \frac{\Phi_0 l^3}{3}; \quad \sigma_B^2 = \Phi_0^2 \int_0^l \int_0^l z_1^2 z_2^2 \sigma_\alpha^2(|z_1 - z_2|) dz_1 dz_2 \qquad [2.10]$$

The relationships given in [2.6]–[2.9] allow us to characterize the redundant force as:

$$X = -\frac{q}{2} \frac{A}{B} \qquad [2.11]$$

showing that it is a non-Gaussian random variable given by the ratio between two known Gaussian variables.

The probability density function (PDF) of X can be obtained in a closed form by applying the PTM to [2.11]. For this purpose, it is necessary to know the Gaussian joint probability density function (JPDF) $p_{AB}(a, b)$ that needs the evaluation of the cross-variance σ_{AB}, besides the already evaluated mean values μ_A and μ_B and variances σ_A^2 and σ_B^2. This cross-variance can be easily obtained starting with the expressions of A and B given in [2.7] i.e.:

$$\sigma_{AB} = \Phi_0^2 \int_0^l \int_0^l z_1^3 z_2^2 \sigma_\alpha^2(|z_1 - z_2|) dz_1 dz_2 \qquad [2.12]$$

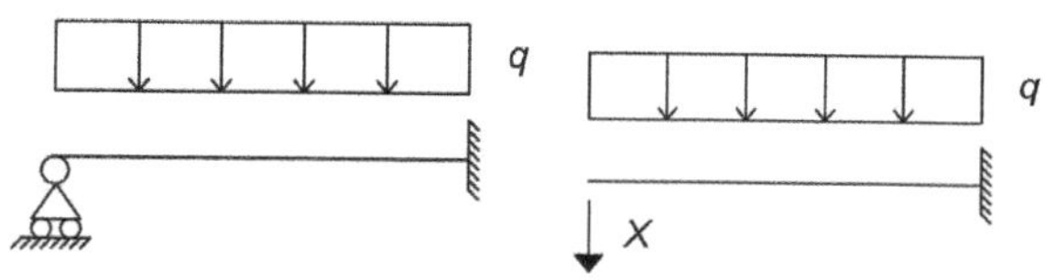

(a) (b)

Figure 2.1. *Example 1: statically indeterminate stochastic beam*

In this way, the expression of the Gaussian JPDF $p_{AB}(a, b)$ can be easily obtained. Once the expression of $\sigma_\alpha^2(|z_1 - z_2|)$ is defined, the variances and cross-variances given in [2.9], [2.10] and [2.12] can be evaluated. Finally, the application of the PTM allows the evaluation of the JPDF $p_{XY}(x, y)$, with Y being an auxiliary random variable that, for example, can be opportunely chosen equal to the Gaussian variable A, obtaining:

$$p_{XY}(x, y) = p_{XA}(x, a) = q\frac{a}{x^2}p_{AB}\left(a, -\frac{q}{2}\frac{a}{x}\right) \qquad [2.13]$$

Finally, the PDF $p_X(x)$ is obtained by the saturation of $p_{XA}(x, a)$ with respect to the variable $Y \equiv A$, i.e.:

$$p_X(x) = \int_{-\infty}^{+\infty} p_{XA}(x, a)da \qquad [2.14]$$

This last evaluation is strongly simplified by the knowledge of the mean and variance of the Gaussian variable A. Once the redundant force has been characterized from a probabilistic point of view, any internal force $S(\bar{z})$ (bending moment or shear force at any abscissa $z = \bar{z}$) can also be characterized. Indeed, it can be written as:

$$S(\bar{z}) = S^{(q)}(\bar{z}) + S^{(X)}(\bar{z}) \qquad [2.15]$$

with $S^{(q)}(\bar{z})$ and $S^{(X)}(\bar{z})$ being the values of the internal forces when the statically determinate structure of Figure 2.1(b) is loaded only by the external force and only by the redundant force, respectively. The first addend is obviously deterministic, while the second one can be rewritten as:

$$S^{(X)}(\bar{z}) = XS^{(1)}(\bar{z}) \qquad [2.16]$$

where $S^{(1)}(\bar{z})$ is the internal force when the redundant force is deterministically unitary. Equations [2.15] and [2.16] imply that $S(\bar{z})$ is linearly dependent on the random variable X. Consequently, its complete probabilistic characterization can be easily defined, once the PDF $p_X(x)$ is known.

The evaluation of the transversal displacements $u(\bar{z})$ is less immediate than the evaluation of $S(\bar{z})$. Indeed, if the relationship analogous to [2.15] is considered for the displacement, then

$$u(\bar{z}) = u^{(q)}(\bar{z}) + u^{(X)}(\bar{z}) \qquad [2.17]$$

In this case, both the variables $u^{(q)}(\bar{z})$ and $u^{(X)}(\bar{z})$ are random. In particular, their expressions are:

$$u^{(q)}(\bar{z}) = \frac{q}{2}\int_{\bar{z}}^{l} z^2(z-\bar{z})\Phi(z)dz = \frac{q}{2}C(\bar{z}), \qquad [2.18]$$

$$u^{(X)}(\bar{z}) = X\int_{\bar{z}}^{l} z(z-\bar{z})\Phi(z)dz = XG(\bar{z}) \qquad [2.19]$$

where $C(\bar{z})$ and $G(\bar{z})$ are two Gaussian random variables having the following mean values and variances:

$$\mu_C(\bar{z}) = \Phi_0\int_{\bar{z}}^{l} z^2(z-\bar{z})dz = \frac{\Phi_0}{12}\left(3l^4 - 4l^3\bar{z} - \bar{z}^4\right), \qquad [2.20]$$

$$\mu_G(\bar{z}) = \Phi_0\int_{\bar{z}}^{l} z(z-\bar{z})dz = \frac{\Phi_0}{6}\left(2l^3 - 3l^2\bar{z} + \bar{z}^3\right), \qquad [2.21]$$

$$\sigma_C^2(\bar{z}) = \Phi_0^2\int_{\bar{z}}^{l}\int_{\bar{z}}^{l} z_1^2 z_2^2(z_1-\bar{z})(z_2-\bar{z})\sigma_\alpha^2(|z_1-z_2|)dz_1dz_2, \qquad [2.22]$$

$$\sigma_G^2(\bar{z}) = \Phi_0^2\int_{\bar{z}}^{l}\int_{\bar{z}}^{l} z_1 z_2(z_1-\bar{z})(z_2-\bar{z})\sigma_\alpha^2(|z_1-z_2|)dz_1dz_2 \qquad [2.23]$$

Rewriting [2.17] in terms of the random variables introduced above, the following expression is obtained:

$$u(\bar{z}) = \frac{q}{2}C(\bar{z}) + XG(\bar{z}) = \frac{q}{2}\left(C(\bar{z}) - \frac{A}{B}G(\bar{z})\right) \qquad [2.24]$$

The random variables appearing in the second member of [2.24] are jointly Gaussian and are characterized by a Gaussian JPDF whose expression is known if, besides the means and the variances of the four quantities, even their cross-variances are obtained. One of them has already been given in [2.12]. The expressions of the other ones are:

$$\sigma_{AC}(\bar{z}) = \Phi_0^2\int_0^{l}\int_{\bar{z}}^{l} z_1^3 z_2^2(z_2-\bar{z})\sigma_\alpha^2(|z_1-z_2|)dz_1dz_2, \qquad [2.25]$$

$$\sigma_{AG}(\bar{z}) = \Phi_0^2\int_0^{l}\int_{\bar{z}}^{l} z_1^3 z_2(z_2-\bar{z})\sigma_\alpha^2(|z_1-z_2|)dz_1dz_2, \qquad [2.26]$$

$$\sigma_{BC}(\bar{z}) = \Phi_0^2 \int_0^l \int_{\bar{z}}^l z_1^2 z_2^2 (z_2 - \bar{z}) \sigma_\alpha^2(|z_1 - z_2|) dz_1 dz_2, \quad [2.27]$$

$$\sigma_{BG}(\bar{z}) = \Phi_0^2 \int_0^l \int_{\bar{z}}^l z_1^2 z_2 (z_2 - \bar{z}) \sigma_\alpha^2(|z_1 - z_2|) dz_1 dz_2, \quad [2.28]$$

$$\sigma_{CG}(\bar{z}) = \Phi_0^2 \int_{\bar{z}}^l \int_{\bar{z}}^l z_1^2 z_2 (z_1 - \bar{z})(z_2 - \bar{z}) \sigma_\alpha^2(|z_1 - z_2|) dz_1 dz_2 \quad [2.29]$$

Once the Gaussian JPDF $p_{ABCG}(a, b, c(\bar{z}), g(\bar{z}))$ is built, the PTM can be applied for the evaluation of the PDF $p_U(u)$, taking into account [2.24] and by considering three auxiliary variables that can be, for example, the same variables A, B and $C(\bar{z})$. Then, the following expression is obtained:

$$p_{ABCu}(a, b, c(\bar{z}), u(\bar{z})) = \frac{q}{2}\frac{a}{b} p_{ABCG}\left(a, b, c(\bar{z}), \frac{b}{a}\left(c(\bar{z}) - \frac{2}{q}u(\bar{z})\right)\right) \quad [2.30]$$

Finally, the PDF $p_U(u)$ is obtained by the saturation of the previous expression with respect to the variables A, B and $C(\bar{z})$. Again, this saturation is simplified by the fact that these are Gaussian known variables.

2.3. Local and non-local randomness

From the previous section, an in-depth view of the integral expressions of the variances, for both the statically and indeterminate stochastic beams, brings up a close similarity of the latter ones with the constitutive equations of the mechanical non-local theory of Eringen (1983). The Eringen non-local integral constitutive equation describes the dependence of the stress at a point on the strain in the rest of the domain through a positive-decaying kernel function (Fernández-Sáez *et al.* 2016). The analogy with the equations giving the statistics of the response stochastic beam is evident. In particular, the analogy between the effect of the kernel function and that of the correlation function is clear.

The flexural deformability $\Phi(z)$ appearing in [2.1] is assumed to be characterized by correlation functions belonging to the classes reported in Table 2.1.

Class	Correlation functions	Parameters
White Noise	$C_{WN}(r) := \delta(r)$	$r \geq 0$
Powered exponential	$C_{EXP}(r) := \exp(\lvert cr\rvert^\alpha)$	$\alpha \in (0, 2]$
Matérn	$C_M(r) := \frac{2^{(\alpha/2)-1}}{\Gamma(\alpha/2)} \lvert cr\rvert^{\alpha/2} K_{\alpha/2}(\lvert cr\rvert)$	$\alpha \in (0, 2]$
Cauchy	$C_C(r) := (1 + \lvert cr\rvert^\alpha)^{\frac{\beta}{\alpha}}$	$\alpha \in (0, 2]$; $\beta > 0$

Table 2.1. *Some parametric classes of correlation functions for a Gaussian process*

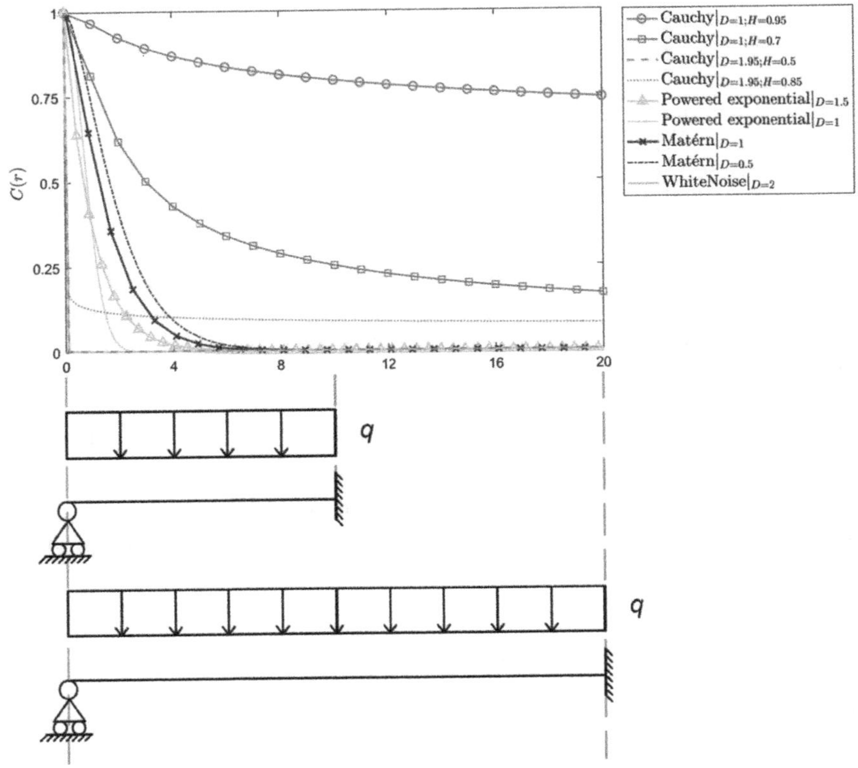

Figure 2.2. *Some parametric classes of correlation functions for a Gaussian process. For a color version of this figure, see www.iste.co.uk/challamel/mechanics3.zip*

It is well known that several properties of an RF can be established through the study of their associated correlation function. The choice of the above correlation functions is motivated by their particular capability to interpret the local and non-local properties of an RF. Relative to these two characteristics, the correlation functions are strictly linked to the local and non-local properties of the RF through two important parameters, respectively: 1) the fractal dimension D reflects the local properties; it is a roughness measure with range $[d, d+1)$, where d is the topological dimension. Since the focus of the present work is on mono-axial beams, $d = 1$ is used; and 2) the Hurst exponent H reflects a long-length dependence in an RF, or, equivalently, a long memory dependence in time series. In particular, the realization of the RF has $D = d + 1 - \alpha/2$ with probability 1 (Gneiting *et al.* 2012), while the RF has a long memory with $H = 1 - \beta/2$ (Gneiting and Schlather 2004). Thus, the parameter α is linked to the fractal dimension, while the parameter β is connected to the Hurst exponent. Therefore, in the field of stochastic mechanics, it is possible to associate the fractal dimension with the local randomness characteristics and the Hurst exponent with the non-local randomness characteristics. The non-local randomness of the flexural deformability RF means that its probabilistic dependence between two different x-points, even if they are far apart from each other, is persistent, i.e. is non-negligible. In comparison, for the local randomness of the flexural deformability RF, if two x-points are taken sufficiently far from each other, then the corresponding RVs are almost uncorrelated. Figure 2.2 illustrates the trend of correlation functions reported in Table 2.1, comparing the different correlation lengths of correlation functions with the different beam lengths that will be investigated in the following sections.

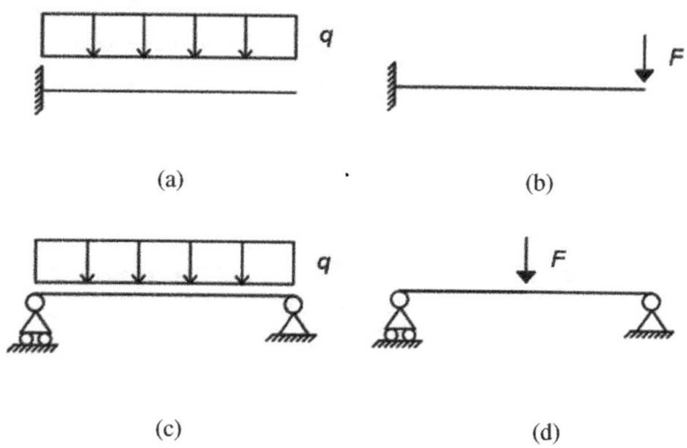

Figure 2.3. *Statically determinate stochastic beams*

In particular, the White Noise, the Powered exponential and the Matérn correlation functions allow the capture of only the local randomness of the flexural deformability characteristics in contrast to the Cauchy class (Gneiting and Schlather 2004) with which it is possible to control at the same time the local and non-local randomness dependence of the flexural deformability characteristic.

In order to investigate the sensitivity of the stochastic response quantities to the local and non-local randomness dependence of the flexural deformability characteristic, in the following sections, some examples of statically determinate and indeterminate stochastic beams, under different conditions of load and constraint, are shown.

2.3.1. *Statically determinate stochastic beams*

The procedure presented in section 2.2.1 has been applied to the stochastic analysis of the statically determinate beams reported in Figure 2.3 for the different types of laws of $\sigma_\alpha^2(|z_1 - z_2|)$ in Table 2.1. Table 2.2 presents a summary of the results; in particular, the estimation of the COV value of the stochastic displacement is interesting, in order to explore for which class of correlation function dispersions of the displacement distribution are more pronounced. In section 2.3.3, some more comments on these results will be reported.

Class	COV $= \sigma/\mu$			
	(a)	(b)	(c)	(d)
Cauchy$\|_{D=1.95;H=0.5}$	0.008	0.008	0.011	0.011
Cauchy$\|_{D=1.95;H=0.85}$	0.086	0.088	0.096	0.095
Powered exponential$\|_{D=1}$	0.131	0.143	0.198	0.196
Powered exponential$\|_{D=1.95}$	0.132	0.144	0.213	0.209
Matérn$\|_{D=1}$	0.161	0.172	0.224	0.223
Matérn$\|_{D=0.5}$	0.176	0.187	0.234	0.233
Cauchy$\|_{D=1;H=0.7}$	0.205	0.211	0.239	0.238
Cauchy$\|_{D=1;H=0.95}$	0.241	0.242	0.248	0.248

Table 2.2. *COV values of the stochastic displacement statically determinate stochastic beams*

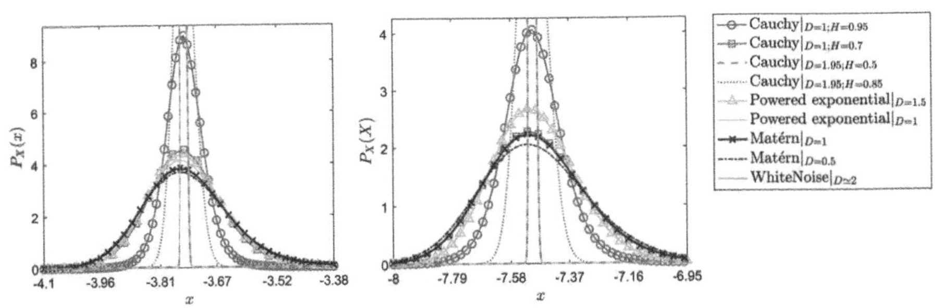

(a) (b)

Figure 2.4. *Example 1: PDF of the redundant force X: (a) $L = 10$; (b) $L = 20$. For a color version of this figure, see www.iste.co.uk/challamel/mechanics3.zip*

2.3.2. *Statically indeterminate stochastic beams*

The stochastic response quantities, which are the redundant force and the displacement of the two examples reported in section 2.2.2, are examined. To not lose generality in the problem under examination, non-dimensional numerical analyses have been evaluated and it is assumed that the average value of the random variables involved is unitary. In Figure 2.4, the PDF of the redundant force X, $p_X(x)$, for the different RF is reported. In particular, the PDF of the stochastic variable X has been examined for two different values of length of the beam, i.e. $L = 10$ (Figure 2.4(a)) and $L = 20$ (Figure 2.4(b)), while Table 2.3 shows the respective COV values of the redundant force. Then, Figure 2.5 shows the PDF of the transversal displacement $u(\bar{z})$, with $\bar{z} = L/2$, for all the assumptions of Gaussian field $\Phi(z)$. Finally, in order to investigate the influence of the abscissa $\bar{z}$ on the stochastic response displacement, in Figures 2.6 and 2.7, the $p_U(u)$ for different values of the abscissa for $L = 10$ and $L = 20$ are shown. Table 2.4 lists all COV values in the investigation on $u(\bar{z})$.

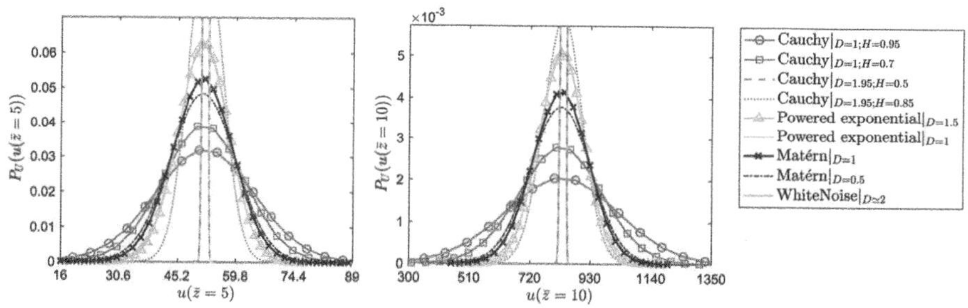

(a) (b)

Figure 2.5. *Example 1: PDF of the transversal displacement $u(\bar{z}) = L/2$: (a) $L = 10$; (b) $L = 20$. For a color version of this figure, see www.iste.co.uk/challamel/mechanics3.zip*

Class	COV $= \sigma/\mu$	
	$L = 10$	$L = 20$
Cauchy$\|_{D=1.95;H=0.5}$	0.001	0.001
Cauchy$\|_{D=1.95;H=0.85}$	0.007	0.006
Powered exponential$\|_{D=1}$	0.016	0.015
Powered exponential$\|_{D=1.95}$	0.026	0.020
Matérn$\|_{D=1}$	0.028	0.021
Matérn$\|_{D=0.5}$	0.030	0.025
Cauchy$\|_{D=1;H=0.7}$	0.033	0.025
Cauchy$\|_{D=1;H=0.95}$	0.039	0.027

Table 2.3. *Example 1: COV values of the redundant force X for $L = 10$ and for $L = 20$*

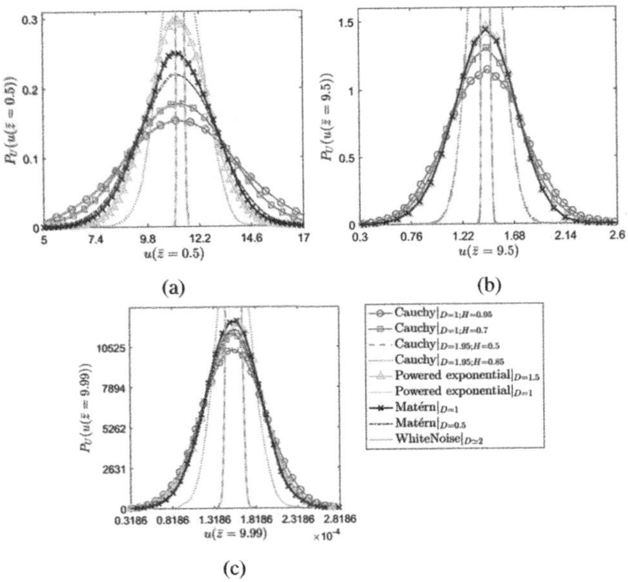

Figure 2.6. *Example 1: PDF of the transversal displacement for* $L = 10$*: (a)* $\bar{z} = 0.5$*; (b)* $\bar{z} = 9.5$*; (c)* $\bar{z} = 9.99$*. For a color version of this figure, see www.iste.co.uk/challamel/mechanics3.zip*

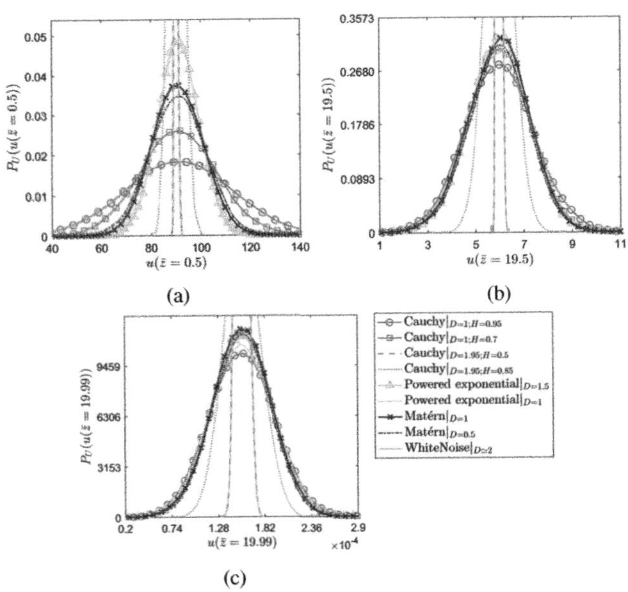

Figure 2.7. *Example 1: PDF of the transversal displacement for* $L = 20$*: (a)* $\bar{z} = 0.5$*; (b)* $\bar{z} = 9.5$*; (c)* $\bar{z} = 9.99$*. For a color version of this figure, see www.iste.co.uk/challamel/mechanics3.zip*

Class	COV $= \sigma/\mu$							
	$L = 10$				$L = 20$			
	$\bar{z} = 0.5$	$\bar{z} = 5$	$\bar{z} = 9.5$	$\bar{z} = 9.99$	$\bar{z} = 0.5$	$\bar{z} = 10$	$\bar{z} = 19.5$	$\bar{z} = 19.99$
Cauchy$\|_{D=1.95;H=0.5}$	0.008	0.013	0.018	0.034	0.009	0.011	0.010	0.098
Cauchy$\|_{D=1.95;H=0.85}$	0.085	0.085	0.098	0.132	0.001	0.081	0.058	0.146
Powered exponential$\|_{D=1}$	0.122	0.124	0.195	0.220	0.113	0.095	0.208	0.224
Powered exponential$\|_{D=1.95}$	0.124	0.126	0.204	0.218	0.116	0.096	0.226	0.250
Matérn$\|_{D=1}$	0.146	0.148	0.210	0.209	0.181	0.117	0.185	0.185
Matérn$\|_{D=0.5}$	0.159	0.161	0.205	0.199	0.169	0.128	0.228	0.192
Cauchy$\|_{D=1;H=0.7}$	0.199	0.197	0.203	0.221	0.145	0.173	0.230	0.220
Cauchy$\|_{D=1;H=0.95}$	0.239	0.234	0.242	0.240	0.236	0.233	0.246	0.266

Table 2.4. *Example 1: COV values of the redundant force X for $L = 10$ and for $L = 20$*

2.3.3. *Comments on the results*

By the analysis of the results of the previous section, the following conclusions can be drawn:

– About the statically determinate stochastic beams, all the output values of the displacement COV of the beams under examination are bounded in a lower and upper bound limit. We obtain the upper limit, i.e. the maximum value of the displacement COV, when the Gaussian RF $\Phi(z)$ follows the Cauchy class characterized by a low value of local randomness ($D = 1$) but at the same time with a high value of non-local randomness ($H = 0.95$). Contrarily, a lower bound limit can be appreciated when $\Phi(z)$ is modeled as an RF totally uncorrelated (White Noise correlation function) with fractal dimension $D = 2$. Analogous results are obtained in the case of the Cauchy class with $D = 1.95$ and $H = 0.5$. Another relevant result about the trend of the displacement variances related to the Cauchy class is that, for fixed values of D, the response variance increases when H increases. Finally, about the response variance, for the Powered exponential and Matérn classes of correlation, no significant difference has been revealed, and the concerning results are confined between the above cited lower and upper bound limits.

– Regarding the statically indeterminate stochastic beams, comparing the results of the transversal displacement at $\bar{z} = L/2$ with the results of the statically determinate examples, it seems that the trend is similar, i.e. an upper bound limit can be observed for a high value of non-local randomness of $\Phi(z)$ and a lower bound limit as the local randomness characteristic increases. However, the same conclusion cannot be made if the redundant force is analyzed. Although for the assumption of RF totally uncorrelated, a relatively low value of COV can be observed, for other laws of $\sigma_\Phi^2(|z_1 - z_2|)$, the trend of the results seems to invert. More dispersed PDFs are obtained for the RF classes that capture only the fractal dimension, i.e. the local randomness. Furthermore, with a focus on the output for the Cauchy class, the variance value of the redundant force increases as the non-local effect (H) decreases. From this last observation, it seems that, in contrast to the response displacement quantity, which is more sensitive to the non-local randomness of $\Phi(z)$, the stochastic internal force response is more sensitive to the local randomness effect. In order to gain deeper insight into the effect of the level of randomness on the type of stochastic response examined, the investigations of the sensitivity of the transversal displacement for different values of the abscissa $\bar{z}$ have been evaluated. It can be appreciated how the tendency of sensitivity to the non-local randomness of the transversal displacement decreases when an abscissa value away from the center of the beam is chosen, so approaching to a major sensitivity to the local randomness, as the redundant force.

2.4. Conclusion

In this chapter, a study on the sensitivity of some random response quantities of stochastic beams to the local and non-local randomness of their mechanical properties of deformability (or stiffness) has been conducted. In particular, it has shown the existence of a relationship between the concept of local and non-local constitutive laws in the Eringen mechanical theory integral constitutive and the concept of local and non-local randomness in the RF describing the deformability of the stochastic uncertain beams. In fact, the role of the kernel function, in the Eringen theory, is played by the correlation function of the homogeneous RF considered in the stochastic beams. The level of the response randomness has been related to the shape of the corresponding PDFs and the values of COVs of the transversal displacements, together with the COVs of the shear internal force and of the bending moment, when the statically indeterminate beams have been examined. From the analysis of the PDFs and the COV values of the response quantities, the following conclusions can be underlined: a) for both statically determinate and indeterminate beams, the stochastic displacements are characterized by COVs which are well ordered, in the sense that the largest COV is related to the RF having the highest values of non-local randomness quantities, while the smallest COV corresponds to the case in which the RF is practically uncorrelated; b) by using the properties of the Cauchy correlation function, referring always to the displacement response, when the coefficient H increases, for fixed value of D, the response PDFs are increasingly dispersed (i.e. the corresponding COVs are higher and higher); c) an opposite trend has been observed about the redundant force, i.e. greater values of COV have been obtained for the RF classes that only capture the fractal dimension. For the Cauchy correlation function, the COV values of the random redundant force increase when the non-local parameter H decreases; and d) the level of sensitivity to the non-local randomness of the transversal displacement depends on the choice of the position . In particular, the randomness level of the transversal displacement is greater in the midpoint of the beam for RF having non-local randomness, while, away from the midpoint, the randomness level of the transversal displacement decreases and is more affected by the local RFs. Definitively, it can be affirmed that the randomness level of both cinematic beam response and, for statically indeterminate beams, static response are strongly affected by the type of correlation function characterizing the RF of the beam deformability. In particular, it is very important to know if this RF has local or non-local properties.

2.5. References

Demmie, P. and Ostoja-Starzewski, M. (2016). Local and nonlocal material models, spatial randomness, and impact loading. *Archive of Applied Mechanics*, 86(1–2), 39–58.

Elishakoff, I., Ren, Y., Shinozuka, M. (1995). Some exact solutions for the bending of beams with spatially stochastic stiffness. *International Journal of Solids and Structures*, 32(16), 2315–2327.

Elishakoff, I., Impollonia, N., Ren, Y. (1999). New exact solutions for randomly loaded beams with stochastic flexibility. *International Journal of Solids and Structures*, 36(16), 2325–2340.

Eringen, A.C. (1983). On differential equations of nonlocal elasticity and solutions of screw dislocation and surface waves. *Journal of Applied Physics*, 54(9), 4703–4710.

Falsone, G. and Laudani, R. (2018). A probability transformation method (PTM) for the dynamic stochastic response of structures with non-Gaussian excitations. *Engineering Computations*, 35(5), 1978–1997.

Falsone, G. and Laudani, R. (2019a). Matching the principal deformation mode method with the probability transformation method for the analysis of uncertain systems. *Engineering Computations*, 118(7), 395–410.

Falsone, G. and Laudani, R. (2019b). Exact response probability density functions of some uncertain structural systems. *Archives of Mechanics*, 71(4–5), 315–336.

Falsone, G. and Settineri, D. (2013a). Explicit solutions for the response probability density function of linear systems subjected to random static loads. *Probabilistic Engineering Mechanics*, 33, 86–94.

Falsone, G. and Settineri, D. (2013b). Explicit solutions for the response probability density function of nonlinear transformations of static random inputs. *Probabilistic Engineering Mechanics*, 33, 79–85.

Fernández-Sáez, J., Zaera, R., Loya, J., Reddy J. (2016). Bending of Euler–Bernoulli beams using Eringen's integral formulation: A paradox resolved. *International Journal of Engineering Science*, 99, 107–116.

Ghavanloo, E., Rafii-Tabar, H., Fazelzadeh, S.A. (2019). Essential concepts from nonlocal elasticity theory. In *Computational Continuum Mechanics of Nanoscopic Structures*, Ghavanloo, E., Rafii-Tabar, H., Fazelzadeh, S.A. (eds). Springer, Cham.

Gneiting, T. and Schlather, M. (2004). Stochastic models that separate fractal dimension and the Hurst effect. *SIAM Review*, 46(2), 269–282.

Gneiting, T., Ševčíková, H., Percival, D.B. (2012). Estimators of fractal dimension: Assessing the roughness of time series and spatial data. *Statistical Science*, 247–277.

Jaw, S. and Chen, C. (1999). Near-wall turbulence modeling using fractal dimensions. *Journal of Engineering Mechanics*, 125, 804–811.

Laudani, R. and Ostoja-Starzewski, M. (2020). Fracture of beams with random field properties: Fractal and Hurst effects. *International Journal of Solids and Structures*, 191, 243–253.

Laudani, R., Dansong, Z., Tarik, F., Porcu, E., Ostoja-Starzewski, M., Chamorro, L. (2020). On streamwise velocity spectra model with fractal and long-memory effects. *Physics of Fluids*. DOI: 10.1063/5.0040453.

Scotti, A., Meneveau, C., Saddoughi, S.G. (1995). Fractal dimension of velocity signals in high-Reynolds-number hydrodynamic turbulence. *Physical Review E*, 51, 5594.

Shen L., Ostoja-Starzewski M., Porcu E. (2015). Harmonic oscillator driven by random processes having fractal and Hurst effects. *Acta Mechanica*, 226(11), 3653–3672.

Turcotte, D.L. (1997). *Fractals and Chaos in Geology and Geophysics*. Cambridge University Press, UK.

Vanmarcke, E. (2010). *Random Fields: Analysis and Synthesis*. World Scientific, New Jersey.

3

On the Applicability of First-Order Approximations for Design Optimization under Uncertainty

3.1. Introduction

In this chapter, different approaches for design optimization under uncertainty are discussed. While the focus is on probabilistic approaches to handle uncertainty, parallels to convex anti-optimization are drawn as well. The most classical probabilistic approaches for uncertainty propagation are the Monte Carlo method of Metropolis and Ulam (1949), and approaches based on a Taylor series approximation like the first-order reliability methods proposed by Cornell (1969) and Hasofer and Lind (1974). Recently, much research has been done on boosting the efficiency of Monte Carlo methods with various surrogate models (see, for example, the overview articles of Sudret *et al.* (2017) or Dey *et al.* (2017), or the work of Balokas *et al.* (2019)).

For design optimization under uncertainty, methods for uncertainty propagation need to be embedded and called (in some cases, multiple times) in each optimization iteration. Hence, Monte Carlo approaches, even with surrogates, become infeasible for applications with many design variables (e.g. non-parametric shape optimization or topology optimization) and/or with many random parameters (e.g. discretized random fields of geometric deviations or spatially varying material parameters). Hence, probabilistic approaches based on Taylor series are still widely used in structural design optimization under uncertainty, for instance for sizing optimization (see Doltsinis and Kang (2004); Guest and Igusa (2008); Jalalpour *et al.* (2011)),

Chapter written by Benedikt KRIEGESMANN.

shape optimization (see Papoutsis-Kiachagias *et al.* (2012); Fragkos *et al.* (2019)) and topology optimization (see Lazarov *et al.* (2012); Kriegesmann and Lüdeker (2019)).

In this chapter, probabilistic approaches based on Taylor series are distinguished by the point of the series expansion. Expansions at the mean input vector are typically used to determine the mean value and variance of the output function (see, for example, Cornell (1969) and Elishakoff *et al.* (1987)). For determining a probability of exceedance (i.e. probability of failure), a Taylor series is expanded at a characteristic point on the *limit state function*. This point is referred to as the *design point*, the *most probable point* (MPP) (as by Haldar and Mahadevan 1999) or as the *most central limit-state point* (as in Ditlevsen and Madsen (1996)), which in general is found by solving an optimization problem.

Design optimization approaches using probabilistic methods can be subdivided into reliability-based design optimization (RBDO) and robust design optimization (RDO). RBDO problems typically have a constraint that restricts a probability of occurrence (e.g. the probability that the maximum stress exceeds an allowed stress level). Therefore, approximations at the limit state function are typically used for this type of problem (see Schuëller and Valdebenito (2010)). In contrast, RDO problems aim to reduce the scatter of the output response by including it in the objective function. Hence, approximations at the mean vector, which provides mean and variance of the output, are often used for this (see Park *et al.* (2006)). For an overview, refer to the review articles of Beyer and Sendhoff (2007), Schuëller and Jensen (2008) or Kanno (2020).

The applicability of first-order reliability methods, i.e. a linear approximation of the objective function (or, respectively, the limit state function) at the MPP, has been discussed extensively, for instance, by Ditlevsen and Madsen (1996) and Dolinski (1982). Measures can be taken to circumvent the approximation error coming from linearization (e.g. Cho (2013)). Hence, RBDO methods using first-order limit state approximations have successfully been applied to problems with highly nonlinear limit state functions and many design variables, such as topology optimization (see Luo *et al.* (2014)).

For RDO problems using approximations at the mean vector, many researchers have emphasized the need for higher-order approaches of the variance (e.g. Fragkos *et al.* (2019)). The computational cost of second-order approaches scales with the number of random parameters, as Asadpoure *et al.* (2011) point out. Even when reducing the number of parameters by *principal component analysis* (also referred to as the *discrete Karhunen–Loève transform*), the computational effort increases by order of magnitude compared to a deterministic optimization. Kriegesmann and Lüdeker (2019), however, showed that by using a first-order probabilistic approach for RDO, the computational effort is less than doubled compared to a deterministic optimization, irrespective of the number of design variables and random parameters. It is therefore worth examining, in detail, in which situations first-order approaches are applicable for RDO, which is presented in what follows.

There are various alternative ways to handle uncertainties, which Elishakoff (1990) divided into *Theory of Probability*, *Fuzzy Sets* and *Anti-Optimization with Set-Theoretical Approaches*. Aside from probabilistic approaches, anti-optimization in convex sets is also discussed here, as has been suggested by Ben-Haim and Elishakoff (1990), which is also based on a Taylor series expansion of the objective function. Also, this approach has already been embedded into topology optimization under uncertainty by Wang *et al.* (2018). Wang *et al.* (2011b) emphasized the similarities between the probability and convexity concepts, and Kriegesmann *et al.* (2012) showed, for the example of composite cylinders, that using a first-order Taylor series for both a probabilistic approach and convex anti-optimization provides nearly the same optimal design, when embedded into a design optimization under uncertainty. This indicates that conclusions of the applicability of first-order probabilistic approaches also hold for convex anti-optimization.

First, a summary is given of the theory of probabilistic approaches and convex anti-optimization based on Taylor series. Then, typical formulations of RDO and RBDO problems are recalled. Finally, by application to typical examples, the applicability of first-order approaches is demonstrated.

3.2. Summary of first- and second-order Taylor series approximations for uncertainty quantification

Consider a function $g(\boldsymbol{x})$, whose input parameters $x_1, ..., x_n$ (summarized in the vector $\boldsymbol{x}$) are subject to scatter, or, in other words, are uncertain. The approaches discussed in this section are based on a Taylor series approximation of g. Haldar and Mahadevan (1999) categorize probabilistic approaches that are based on Taylor series by the order of the polynomial, i.e. first-order versus second-order approaches. Here, a different categorization is chosen based on the location at which the Taylor series is expanded, namely at the mean vector (section 3.2.1) or at the so-called *most probable point* (section 3.2.2). In the context of convex anti-optimization, the same approximation of g may be used, which leads to similar types of approaches. The similarities of Taylor series-based probabilistic approaches and convex anti-optimization are discussed in section 3.2.4.

The focus of this chapter is on the approaches given in section 3.2.1. The other approaches are also discussed to show the similarities as well as to clearly distinguish between the approaches because of the sometimes non-uniform denomination in the literature.

3.2.1. *Approximations of stochastic moments*

In this section, the first-order second-moment (FOSM) method and its higher-order equivalents are described. The method is also referred to as the *method of moments* (e.g. by Papoutsis-Kiachagias *et al.* (2012)). It is based on a Taylor series expansion g at the mean vector $\boldsymbol{\mu}$.

$$\begin{aligned} g(\boldsymbol{x}) &= g(\boldsymbol{\mu}) + \sum_{i=1}^{n} \frac{\partial g(\boldsymbol{\mu})}{\partial x_i}(x_i - \mu_i) \\ &+ \frac{1}{2}\sum_{i=1}^{n}\sum_{j=1}^{n} \frac{\partial^2 g(\boldsymbol{\mu})}{\partial x_i\, \partial x_j}(x_i - \mu_i)(x_j - \mu_j) + \ldots \end{aligned} \qquad [3.1]$$

The approach does not aim to determine the full stochastic distribution of g, but only its stochastic moments, i.e. the mean value μ_g and variance σ_g, by

$$\begin{aligned} \mu_g &= \int_{-\infty}^{\infty} g(\boldsymbol{x})\, f_{\boldsymbol{X}}(\boldsymbol{x})\ d\boldsymbol{x} \\ \sigma_g^2 &= \int_{-\infty}^{\infty} \left[g(\boldsymbol{x}) - \mu_g\right]^2 f_{\boldsymbol{X}}(\boldsymbol{x})\ d\boldsymbol{x} \end{aligned} \qquad [3.2]$$

Using the Taylor series [3.1] up to the first-order terms and inserting it into [3.2] yields

$$\mu_g \approx g(\boldsymbol{\mu}) \qquad [3.3]$$

$$\sigma_g^2 \approx \sum_{i=1}^{n}\sum_{j=1}^{n} \frac{\partial g(\boldsymbol{\mu})}{\partial x_i}\frac{\partial g(\boldsymbol{\mu})}{\partial x_j}\mathrm{cov}(X_i, X_j) \qquad [3.4]$$

This approximation contains up to second-moment information of the random input vector $\boldsymbol{X}$. A second-order Taylor series yields approximations which contain up to fourth-order moments when used for determining the variance, which is referred to as the second-order fourth-moment (SOFM) by the author (e.g. in Kriegesmann and Lüdeker (2019)) to differentiate it from approaches that intentionally restrict the input moments taken into account as the second-order third-moment approach by Hong *et al.* (1999). When also the skewness of g is determined, up to sixth-order moments are used (see Kriegesmann *et al.* 2011).

The random vector $\boldsymbol{X}$ can be transformed to a vector $\boldsymbol{Z}$ with uncorrelated entries by

$$\boldsymbol{X} = \boldsymbol{\Sigma}_{\boldsymbol{Z}}^{\frac{1}{2}} + \boldsymbol{\mu}_{\boldsymbol{X}} \quad \Leftrightarrow \quad \boldsymbol{Z} = \boldsymbol{\Sigma}_{\boldsymbol{X}}^{-\frac{1}{2}}(\boldsymbol{X} - \boldsymbol{\mu}_{\boldsymbol{X}}) \qquad [3.5]$$

Here, Σ_X is the covariance matrix and μ_X the mean vector of X. When determining the root of Σ_X by spectral decomposition, the transformation is equivalent to a principal component analysis (PCA – see, for example, Kriegesmann (2012)). Equation [3.4] then simplifies to

$$\sigma_g^2 \approx \sum_{i=1}^{n} \left(\frac{\partial g(\mu)}{\partial z_i} \right)^2 \operatorname{var}(Z_i) \qquad [3.6]$$

The perturbation technique, for instance as used by Lazarov *et al.* (2012), yields similar governing aligns for the mean and variance of g, but it is motivated by a Taylor series expansion of the static equilibrium $\boldsymbol{K}\,\boldsymbol{u} = \boldsymbol{f}$.

3.2.2. Probabilistic lower bound approximation

The following class of approaches determines the probability $F_g(\bar{g})$ that the objective function $g(\boldsymbol{x})$ takes values below or equal to $\bar{g}$. In other words, instead of determining stochastic moments as in the previous section, the cumulative distribution $F_g(g(\boldsymbol{x}))$ is evaluated for one specific value $\bar{g}$. In this context, the function $g_{LS} = g(x) - \bar{g}$ is referred to as the *limit state function*. Given that g is linear, $g_{LS} = 0$ is a hyperplane.

Figure 3.1. *Principle of first-order reliability method, original space (left) and transformed space (right). For a color version of this figure, see www.iste.co.uk/challamel/mechanics3.zip*

The probability that $\bar{g}$ is not exceeded is determined by integrating the probability density $f_X(\boldsymbol{x})$ over the space of $\boldsymbol{x}$ in which $g(\boldsymbol{x}) \leq \bar{g}$ (see Figure 3.1, left), i.e.

$$F_g(\bar{g}) = \int_{\boldsymbol{x}:g(\boldsymbol{x})\leq\bar{g}} f_X(\boldsymbol{x}) d\boldsymbol{x} \qquad [3.7]$$

Alternatively, the integral can be solved in the space of a random input vector U (see Figure 3.1, right), which has zero mean and the identity matrix as the covariance matrix (i.e. variances equal 1 and all covariances equal 0). The transformation from a random parameter X with CDF $F_X(x)$ to a standard Gaussian parameter U with the CDF Φ is given by the Rosenblatt transformation. Correlated parameters can be transformed into uncorrelated parameters with [3.5]. Given a linear objective function g, $F_g(\bar{g})$ can be determined by

$$F_g(\bar{g}) = \Phi(-\beta) \tag{3.8}$$

The point on the limit state function closest to the origin in the $\boldsymbol{U}$-space is referred to as the *most probable point* (MPP). The distance equals the *reliability index* β. For nonlinear problems, the MPP needs to be determined by solving the optimization problem

$$\begin{aligned} &\min_{\boldsymbol{u}} \sqrt{\boldsymbol{u}^T \boldsymbol{u}} \\ &\text{subject to } g(\boldsymbol{u}) = \bar{g} \end{aligned} \tag{3.9}$$

If g and hence also g_{LS} are nonlinear (see the dashed lines in Figure 3.1), assuming g_{LS} to be a hyperplane is a first-order approximation of g at the MPP. Second-order approaches take into account the curvature of g_{LS} at the MPP. For further details, see, for example, Haldar and Mahadevan (1999).

It is often necessary to solve the problem inversely, i.e. determine the value $\bar{g}$ that is associated with a prescribed cumulative probability $F_g(\bar{g})$. This prescribed probability can be expressed as a prescribed reliability index β. If the type of the stochastic distribution of g is known (for instance, because $f_{\boldsymbol{X}}$ is a Gauss distribution and g is linear, and hence, f_g is Gaussian), it is sufficient to determine the stochastic moments of f_g. This can, for instance, be done with the methods presented in section 3.2.1. The design value $\bar{g}$ can then be expressed as

$$\bar{g} = \mu_g + b\sigma_g \tag{3.10}$$

In the case of the Gauss distribution of g, $b = \beta$.

3.2.3. *Convex anti-optimization*

Given a set of measurement data of a vector $\boldsymbol{x}$, the first step of the concept proposed by Ben-Haim and Elishakoff (1990) is to determine a domain of $\boldsymbol{x}$, which is described as a hyper-ellipsoid. The smallest hyper-ellipsoid is referred to as a minimum volume enclosing hyper-ellipsoid (MVEE), which can, for instance, be determined using a genetic algorithm as in Elishakoff *et al.* (2012). The MVEE is characterized by its center $\boldsymbol{x}_c$, the orientations and the lengths s_i of the semi-axes.

The objective function $g(\boldsymbol{x})$ is approximated by a Taylor series that is expended at the center $\boldsymbol{x}_c$ of the MVEE in terms of coordinates $\boldsymbol{\xi}_i$ which are parallel to the semi-axes of the MVEE.

$$g(\boldsymbol{x}_c + \boldsymbol{\xi}) = g(\boldsymbol{x}_c) + \sum_{i=1}^{n} \frac{\partial g(\boldsymbol{x}_c)}{\partial \xi_i} \xi_i + \frac{1}{2} \sum_{i=1}^{n} \sum_{j=1}^{n} \frac{\partial^2 g(\boldsymbol{x}_c)}{\partial \xi_i \, \partial \xi_j} \xi_i \xi_j + \dots \quad [3.11]$$

The convexity of the MVEE allows us to directly determine the minimum value of g inside the MVEE. Using a first-order approximation, the minimum value is given by

$$g_{\min} = g(\boldsymbol{x}_c) - \sqrt{\sum_{i=1}^{n} s_i^2 \left(\frac{\partial g(\boldsymbol{x}_c)}{\partial \xi_i} \right)^2} \quad [3.12]$$

Of course, a hyper-ellipsoid is not the only possible convex domain of uncertain parameters. Another widely used domain is a hyper-rectangle, which is obtained when considering intervals for all parameters. Also, for a rectangular domain, the minimum volume enclosing hyper-rectangle can be fitted to given data (as for instance in Elishakoff *et al.* (2012)). Assuming a monotonic objective function, its extremal values are found at the vertices of the hyper-rectangle. The computational cost therefore equals 2^n function evaluations to check all vertices. Fujita and Takewaki (2011) reduced this computational cost to $2n$ function evaluations, even for non-monotonic objective functions, by using a second-order Taylor series.

In general and for arbitrarily shaped domains, the minimum of the objective function can be found by employing optimization algorithms. Then, the concept of convex anti-optimization is more similar to approaches summarized in section 3.2.2. However, in the following, the approach based on the Taylor series at the center of an hyper-ellipsoid is considered.

3.2.4. Correlation of probabilistic approaches and convex anti-optimization

When using equation [3.10] in combination with the FOSM method, i.e. equations [3.3] and [3.6], the probabilistic lower bound can be written as

$$\bar{g} = \mu_g - b\,\sigma_g = g(\boldsymbol{\mu}) - b \sqrt{\sum_{i=1}^{n} \left(\frac{\partial g(\boldsymbol{\mu})}{\partial z_i} \right)^2 \sigma_{Z_i}^2} \quad [3.13]$$

Equation [3.13] is very similar to equation [3.12], whereupon the location at which the g and its gradient are evaluated is different, the partial derivatives refer to different parameters and the contribution of the derivatives is scaled with different parameters, namely s_i and $b\sigma_{Z_i}$. Still, both approaches tend to provide similar results, which will be explained with the following example. Figure 3.2 shows a set of data points enclosed by the MVEE. The directions ξ_i of the semi-axes are shown as well as the directions z_i of the principal components. The center $\boldsymbol{x}_c$ of the MVEE is close to the mean vector $\boldsymbol{\mu}$, the direction of the largest semi-axis ξ_1 is similar to the direction of the first principal component z_1 and the same holds for ξ_2 and z_2[1]. Of course, this similarity is not valid in general, but it is reasonable to assume that it is typical. The remaining difference is the "scale factors" s_i and $b\sigma_{Z_i}$. It has been demonstrated in Kriegesmann *et al.* (2012) that these "scale factors" do not affect the optimal design that is found in a design optimization under uncertainty.

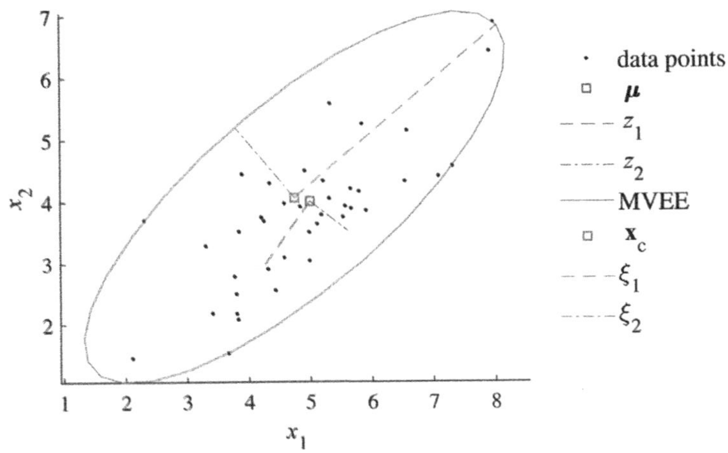

Figure 3.2. *Two-dimensional example of a set of data points, their mean vector $\boldsymbol{\mu}$, principal components z_1 and z_2 and minimum volume enclosing hyper-ellipsoid (MVEE), characterized by its center $\boldsymbol{x}_c$ and semi-axes ξ_1 and ξ_2. For a color version of this figure, see www.iste.co.uk/challamel/mechanics3.zip*

3.3. Design optimization under uncertainty

In the following section, $\boldsymbol{x}$ is still the vector of uncertainties and $\boldsymbol{y}$ is the design vector.

3.3.1. *Robust design optimization*

RDO problems considered here aim to reduce the variability of a probabilistic objective function g (for a classification of RDO problems, see, for example,

1 The sign of the direction of the semi-axes and the principal components are not uniquely defined; therefore, in the example, they happen to be opposite. However, this does not affect the conclusion, as the derivatives are squared in equations [3.12] and [3.13].

Kanno (2020)). A typical optimization problem is to reduce the mean value and, at the same time, the standard deviation of g, i.e.

$$\begin{aligned}&\min_{\boldsymbol{y}} \mu_g(\boldsymbol{y}) + \kappa\,\sigma_g(\boldsymbol{y})\\ &\quad\text{s.t. } \boldsymbol{c}_{ne}(\boldsymbol{y}) \leq \mathbf{0}\\ &\qquad\quad \boldsymbol{c}_{eq}(\boldsymbol{y}) = \mathbf{0}\end{aligned} \qquad [3.14]$$

Here, $\boldsymbol{c}_{ne}$ and $\boldsymbol{c}_{eq}$ represent the generic non-equality and equality constraints. Since only stochastic moments of g have to be evaluated, this class of problem is often tackled with approaches based on Taylor expansions at the mean vector $\boldsymbol{\mu}$, which are discussed in section 3.2.1 (see, for instance, the works of Doltsinis and Kang (2004), Asadpoure *et al.* (2011), Lazarov *et al.* (2012), Fragkos *et al.* (2019) and Kriegesmann and Lüdeker (2019)).

The factor κ is a weighting factor of this multi-objective optimization problem. However, as discussed at the end of section 3.2.2, the objective function of [3.14] can also be considered as a design value $\bar{g}$ (compare align [3.10]), which helps to choose a reasonable value for κ.

3.3.2. *Reliability-based design optimization*

RBDO problems typically minimize some cost function c subject to a probabilistic constraint, i.e.

$$\begin{aligned}&\min_{\boldsymbol{y}} c(\boldsymbol{y})\\ &\text{s.t. } \boldsymbol{c}_{ne}(\boldsymbol{y}) \leq \mathbf{0}\\ &\quad\ \ \boldsymbol{c}_{eq}(\boldsymbol{y}) = \mathbf{0}\\ &P(g(\boldsymbol{X}) \leq \bar{g}) - P_{allow} \leq 0\end{aligned} \qquad [3.15]$$

Again, $\boldsymbol{c}_{ne}$ and $\boldsymbol{c}_{eq}$ represent the generic non-equality and equality constraints. The probability P that $g(\boldsymbol{X})$ does not exceed the value $\bar{g}$ is restricted to be less than or equal to P_{allow}. This requires evaluating the cumulative frequency $F_g(\bar{g})$, and therefore, this class of problem is typically tackled with the approaches summarized in section 3.2.2 (refer also, for instance, to the review article of Schuëller and Valdebenito (2010)).

When using equation [3.10] to determine a lower or upper quantile $\bar{g}$, the probabilistic constraint can be written as

$$\mu_g + b\sigma_g - \bar{g} \leq 0 \qquad [3.16]$$

(The sign of the constraint depends on whether an upper or lower quantile is considered.) Now, the constraint is similar to the objective function of equation [3.14]. And the problem may be solved with the approaches discussed in section 3.2.1.

3.3.3. *Optimization with convex anti-optimization*

Using convex anti-optimization for design optimization under uncertainty can be used to maximize the worst-case value of an objective (for instance, the buckling load as considered by Kriegesmann *et al.* (2012)), or in order to impose constraints on worst-case values, as done by Wang *et al.* (2018). In both cases, equation [3.12] may be used. With the similarities discussed in section 3.2.4, design optimization using a first-order convex anti-optimization will lead to the same problems as RDO with a first-order probabilistic approach.

3.4. Numerical examples

The simple numerical examples in this section demonstrate the applicability of first-order approximations for different random parameters. While section 3.4.1 shows a probabilistic analysis, sections 3.4.2 and 3.4.3 show RDO examples.

3.4.1. *Imperfect von Mises truss analysis*

The von Mises truss, shown in Figure 3.3, is considered to have random geometric imperfections. Thus, the maximum load scatters around the bifurcation point. The length of one bar was chosen to equal 10 and the angle between one bar and the horizontal axis equals $\pi/6$. The axial stiffness equals $EA = 96800$ and the bending stiffness equals $EI = 390.4$. Each bar is discretized with 10 elements, and nonlinear finite element analyses were conducted to determine the buckling load of the truss.

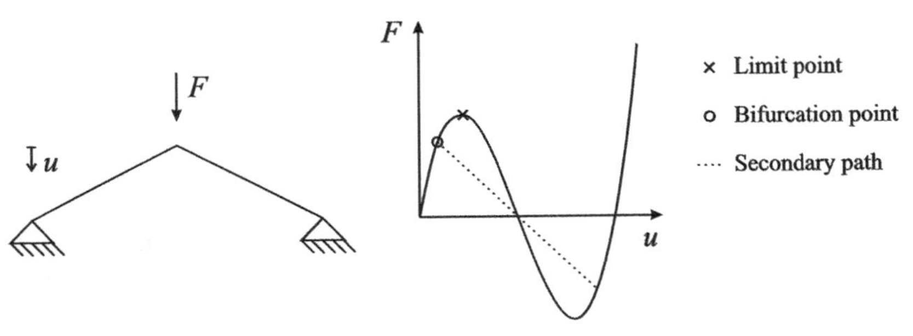

Figure 3.3. *von Mises truss (left) and associated load–displacement curve (right)*

The geometric deviation perpendicular to the beam axis is chosen as a homogeneous random field with Gaussian auto-correlation function $r(\Delta x) = exp(-\Delta x^2/l_c^2)$, a correlation length of $l_c = 2$ and a standard deviation of $\sigma = 0.1$. To simplify the analysis, symmetry was enforced, i.e. both bars have the same "mirrored" imperfection. In one case, the mean value was assumed to equal 0, i.e. the perfect structure equals the mean structure. In a second case, a realization of the random field was used as mean imperfection, which equals μ = (-0.1750, -0.1539,-0.1626,-0.1119,0.0266,0.1649,0.2674,0.1887,0.0086,0.0086,-0.0401). Each entry corresponds to the perpendicular deviation at one finite element node in an equidistant mesh.

The von Mises truss was analyzed probabilistically with Monte Carlo simulations, the FOSM method, and also a higher-order approach, the second-order fourth-moment (SOFM) method (see, for example, Kriegesmann and Lüdeker (2019)). The derivatives are approximated using central finite differences. The results are summarized in Table 3.1. When the mean imperfection equals the perfect structure, the first- and second-order approaches provide unrealistic results, far off the Monte Carlo simulation. For the non-zero mean imperfection, however, the results of the first-order approximation match the Monte Carlo simulation.

	Zero mean		Non-zero mean	
Approach	μ_{BL}	σ_{BL}	μ_{BL}	σ_{BL}
FOSM	37.35	30.8	22.1	2.52
SOFM	$-3.8 \cdot 10^3$	$2.3 \cdot 10^3$	22.2	2.54
Monte Carlo	25.3	7.45	22.3	2.56

Table 3.1. *Mean values and standard deviations of the buckling load of the von Mises truss with random imperfections*

For the zero mean imperfection, any deviation from the mean leads to a drop of the buckling load. Therefore, the FOSM method overestimates the mean buckling load. The imperfections applied for the finite difference steps may cause a change to the qualitative behavior, and the buckling load moves from the limit point to below the bifurcation point. This leads to bad approximations of the derivatives, resulting in unreasonable values for the standard deviation obtained by FOSM, as well as for the mean and standard deviation obtained by SOFM.

For the non-zero mean imperfection, however, the behavior of the buckling load with respect to geometric deviations is smoother and monotonic, which results in the good agreement of all approaches. In real applications like cylindrical shell buckling, a non-zero imperfection is more realistic (see, for example, Arbocz and Hol (1991), Chryssanthopoulos *et al.* (1991), Chryssanthopoulos and Poggi (1995), Kriegesmann *et al.* (2010), Kriegesmann *et al.* (2011) and Schillo *et al.* (2015)), which justifies the applicability of first-order approaches for such cases.

3.4.2. *Three-bar truss optimization*

The next example is the three-bar truss shown in Figure 3.4, which is the same as in Kriegesmann (2020) and similar to an example from Lee and Park (2001). The nominal cross-sections of the bars $\bar{a}_k$ are considered as design variables. All the other measures are kept constant with $H = 40$, $B = 30$, Young's modulus of $E = 1000$ and a vertical of $P_v = 10$.

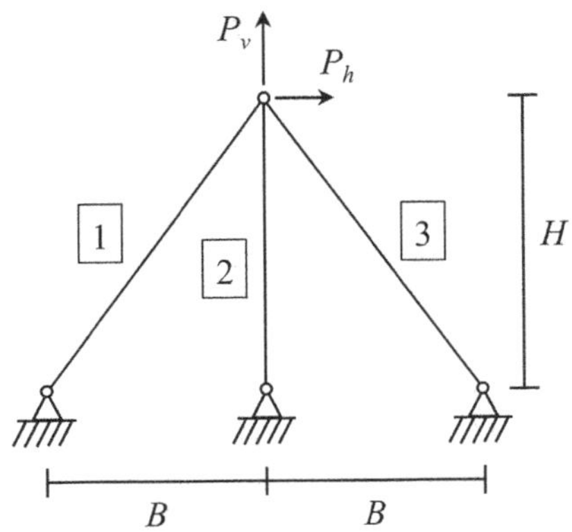

Figure 3.4. *Three-bar truss example*

The optimization objective is to minimize the compliance $c = \boldsymbol{u}^T \boldsymbol{f}$ in the deterministic optimization. In RDO, the sum of the mean and standard deviation of the compliance is minimized with a weight factor of $\kappa = 5$. The sum of cross-sections is constrained to equal 3.0, and the cross-sections $\bar{a}_k$ may only take positive values. The RDO problem is given by

$$
\begin{aligned}
&\min_{\bar{\boldsymbol{a}}} \mu_c + \kappa \, \sigma_c \\
&\text{subject to} \\
&\bar{a}_1 + \bar{a}_2 + \bar{a}_3 = 3.0 \\
&\bar{a}_k \geq 0 \\
&\boldsymbol{K}\,\boldsymbol{u} = \boldsymbol{f}
\end{aligned}
\qquad [3.17]
$$

3.4.2.1. *Cross-section as design and random parameters – first-order works*

First, the cross-sections are considered as random parameters. Since then, design variables and random parameters refer to the same physical property, the cross-section is split into the nominal cross-section $\bar{a}_k$, which is the design variable, and the scattering part $\tilde{a}_k$, which is the random parameter

$$a_k = \bar{a}_k + \tilde{a}_k \qquad [3.18]$$

The random parameters $\tilde{A}_k$ are chosen to scatter uniformly in the interval $[-0.15\bar{a}_k^2, 0.25\bar{a}_k^2]$. A possible motivation is a manufacturing tolerance for the width and height of a rectangular cross-section, which is defined relative to the nominal value. The random parameters are assumed to be independently distributed. Note that the distributions of random parameters $\tilde{A}_k$ are dependent on the design variables $\bar{a}_k$. This requires some special consideration when determining the gradient of the optimization problem, which is discussed in detail in Kriegesmann (2020).

In this section, the horizontal load is constant and equals $P_h = 0$. The results of a deterministic optimization and an RDO are given in Table 3.2. Unsurprisingly, the deterministic optimization yields a design which consists of the central bar only. The RDO using the FOSM method distributes the material among the three bars, leading to a lower standard deviation of the compliance. The optimized results are in both cases analyzed with the FOSM method and Monte Carlo simulations with 100,000 realizations. The deviation of FOSM and Monte Carlo is small enough so that the RDO still provides a much more robust design than the deterministic optimization (at the cost of an increased mean compliance). Hence, the FOSM method may well be used for RDO in this case, even though design variables and random parameters refer to the same physical property.

Optimization approach	Optimal design parameters			FOSM		Monte Carlo	
	$\bar{a}_1$	$\bar{a}_2$	$\bar{a}_3$	μ_c	σ_c	μ_c	σ_c
Deterministic	$0.8 \cdot 10^{-8}$	3.5	$0.8 \cdot 10^{-8}$	0.973	0.335	1.12	0.459
RDO with FOSM	0.90	1.25	0.90	1.74	0.147	1.76	0.150

Table 3.2. *Results of optimizations of the three-bar truss example with random cross-sections*

3.4.2.2. *Load as a random variable – first-order fails*

Next, the horizontal load P_h is considered as a random parameter, which is uniformly distributed in the interval (-2,2). The cross-sections now equal the nominal cross-sections, i.e. $a_k = \bar{a}_k$. Apart from the FOSM method, the SOFM approach is also applied.

Unsurprisingly, the results presented in Table 3.3 show that the deterministically optimized design is extremely sensitive to variations of the horizontal load. For this design, the FOSM method determines a standard deviation of approximately zero.

The reason for this well-known phenomenon is that the derivative of the objective function with respect to the random parameter is zero for this configuration, i.e.

$$\frac{\partial g(\boldsymbol{\mu})}{\partial x} = \frac{\partial c(0)}{\partial P_v} = 0 \qquad [3.19]$$

From equation [3.4], it can be seen that the variance estimated by FOSM then equals zero. Deterministic optimization and RDO using FOSM therefore provide the same result. In contrast, using a second-order Taylor expansion provides accurate results and hence, embedding it into RDO yields a robust design; see Figure 3.5.

Optimization approach	Optimal design parameters $\bar{a}_1$	$\bar{a}_2$	$\bar{a}_3$	FOSM μ_c	σ_c	SOFM μ_c	σ_c	Monte Carlo μ_c	σ_c
Deterministic	$1.1 \cdot 10^{-6}$	3.5	$1.1 \cdot 10^{-6}$	1.14	$1.0 \cdot 10^{-9}$	87161	77958	87186	77910
RDO with FOSM	0.009	3.48	0.009	1.15	$1.7 \cdot 10^{-5}$	11.36	9.14	11.37	9.13
RDO with SOFM	0.716	1.71	0.716	1.64	$1.7 \cdot 10^{-8}$	1.77	0.116	1.77	0.116

Table 3.3. *Results of optimizations of the three-bar truss example with random horizontal load*

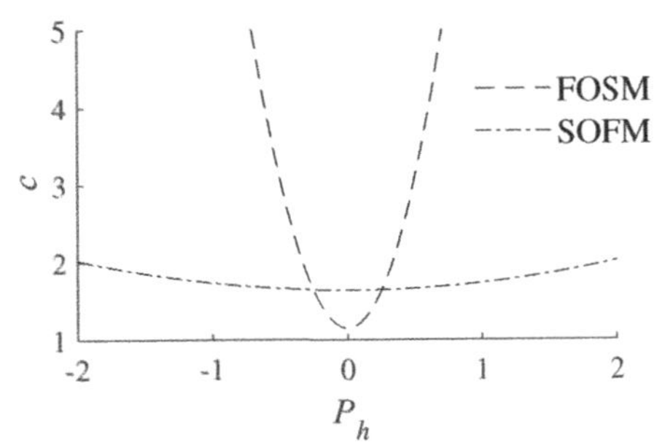

Figure 3.5. *Compliance over horizontal load P_h for the optimized design using the FOSM and the SOFM methods*

The same conclusion holds for optimization under uncertainty using convex anti-optimization. Using a first-order approach equation [3.12], the lower bound is found at the center point $\boldsymbol{x}_c$ because the derivatives vanish.

3.4.3. *Topology optimization*

In order to demonstrate the implications in the presence of many design parameters, two examples of topology optimization are considered in the following section. These examples demonstrate the huge computational benefit of using a first-order probabilistic approach, but they also show that the same problems occur as for the previous example, when considering a random load.

In both cases, a design space of 150 x 50 is used with an element size of 1. The number of design parameters equals the number of elements, which equals $150 \times 50 = 7500$. Linear finite element analyses are carried out using linear 2D elements with full integration and plane stress.

The optimization problem considered is given by

$$\begin{aligned}
&\min_{\boldsymbol{\rho}} \mu_c(\boldsymbol{\rho}) + \kappa\, \sigma_c(\boldsymbol{\rho}) \\
&\text{subject to} \\
&\frac{V(\boldsymbol{\rho})}{V_0} \leq v \\
&0 \leq \rho_e \leq 1 \\
&\boldsymbol{K}\boldsymbol{u} = \boldsymbol{f}
\end{aligned} \tag{3.20}$$

Similarly, as in section 3.4.2, the mean and standard deviation of the compliance are considered as the objective function of the robust topology optimization (RTO). For the deterministic optimization, the compliance itself is the objective function. The design variables are the pseudo-element densities ρ_e, summarized in the design vector $\boldsymbol{\rho}$. The volume V is constrained to be a fraction v of the design space volume V_0.

The simplified isotropic material with penalization (SIMP) approach of Bendsøe (1989) is used with a penalization factor of $p = 3$. The design variables are filtered using the density filter of Guest *et al.* (2004) and projected by the Heaviside approximation of Wang *et al.* (2011a) with projection parameters of $\eta = 0.5$ and $\beta = 10$. For all examples, the method of moving asymptotes of Svanberg (1987) is used with some modifications suggested by Li and Khandelwal (2014). For a detailed overview of the topology optimization settings used, refer to Kriegesmann and Lüdeker (2019).

3.4.3.1. *Load as a random variable – first-order fails*

The first topology optimization problem is shown in Figure 3.6, where the horizontal force is random uniformly distributed in the interval (-2,2). Similar examples have been considered in many works such as Maute and Frangopol (2003), Gournay *et al.* (2008), Luo *et al.* (2014) and Carrasco *et al.* (2015).

The parameters used for this example are Young's modulus of $E = 1$, Poisson's ratio of $\nu = 0.3$, a volume fraction of $v = 10\%$ and a filter radius of $R = 4$. The deterministic vertical load component equals $F_y = 1$ and the horizontal load F_x has a mean value of $\mu_{F_y} = 0$ and a standard deviation of $\sigma_{F_y} = 1/3$. The RTO problem [3.20] is solved with $\kappa = 3$.

The results of deterministic optimization and RTO are shown in Figure 3.7. For the RTO result in Figure 3.7 (right) the SOFM method is used, whereas an optimization with Monte Carlo provided almost the same result. Using the FOSM method within the RTO provides almost the same result as the deterministic optimization. The reason is the same as for the three-bar truss example: the derivative of the compliance w.r.t. the random load becomes zero, and hence, the FOSM approach estimates a variance of zero. The mean values and standard deviation of the compliance given in Table 3.4 are determined by Monte Carlo simulation, if not stated otherwise.

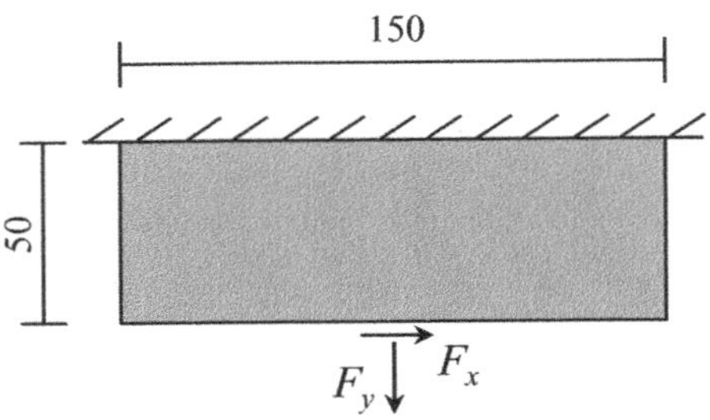

Figure 3.6. *Design space and load of the tension bar example*

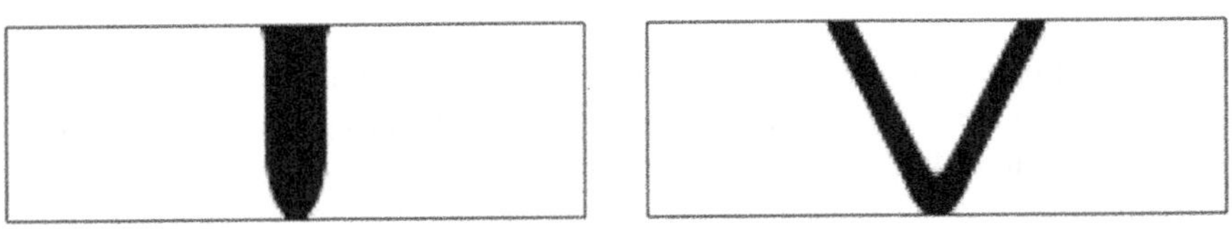

Figure 3.7. *Result of deterministic (left) and robust (right) topology optimization of the tension bar with random load*

Approach	μ_c	σ_c
Deterministic	23.3	33.9
Monte Carlo	11.4	4.4
FOSM	14.0*, 35.3	0.5*, 39.7
SOFM	10.0**, 10.5	3.2**, 5.0

*Determined by the FOSM, **determined by the SOFM.

Table 3.4. *Mean value and standard deviation of the compliance of the design obtained by different RTO approaches*

3.4.3.2. *Young's modulus as a random field – first-order works*

In the next example, the cantilever beam shown in Figure 3.8 is considered with an applied load of $F = 1$.

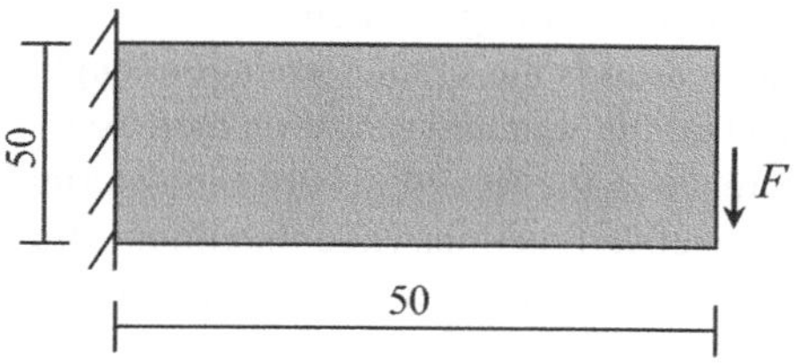

Figure 3.8. *Design space and load of the cantilever beam example*

Young's modulus is considered as a spatially varying random parameter. This random field is discretized with the same mesh as the finite element mesh. The number of random variables considered hence equals 7500. The mean and standard deviation of Young's modulus equal $\mu_E = 10$ and $\sigma_E = 2.5$. The random field is assumed to be homogeneous, and the Gaussian correlation function is chosen with a correlation length of $l_c = 2$

$$C(\Delta \boldsymbol{x}) = \exp\left(-\frac{\Delta \boldsymbol{x}^2}{l_c^2}\right) \qquad [3.21]$$

For the topology optimization, a filter radius of $R = 2$ is used and the prescribed volume fraction equals $v = 0.5$. The weight factor of the RTO is chosen as $\kappa = 5$.

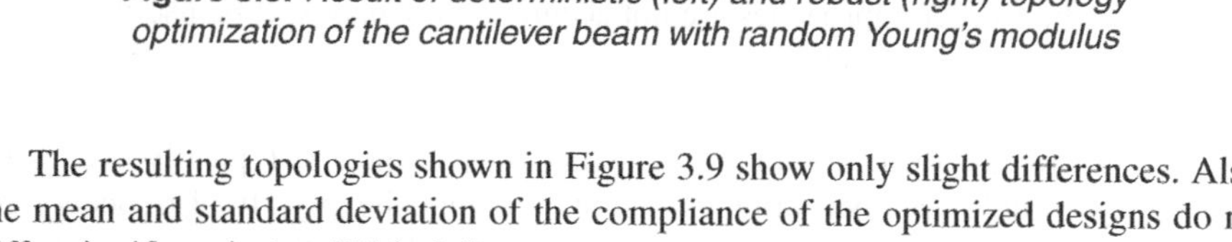

Figure 3.9. *Result of deterministic (left) and robust (right) topology optimization of the cantilever beam with random Young's modulus*

The resulting topologies shown in Figure 3.9 show only slight differences. Also, the mean and standard deviation of the compliance of the optimized designs do not differ significantly (see Table 3.5).

For validation, Monte Carlo simulations of the optimized designs were performed with 10,000 realizations. Here, Young's modulus was assumed to follow the Weibull distribution in order to avoid negative values, but the mean and covariance of Young's modulus remained the same. Again, the Monte Carlo results deviate from the FOSM results, but the FOSM provides the same tendency and therefore a more robust design.

In this example, the standard deviations are very small, keeping in mind that the assumed standard deviation of Young's modulus (25%) is very high compared to most real materials. The reason is the small correlation length. For many geometries, the authors experienced that the standard deviation increases as the correlation length increases, but the difference in the topology between deterministic optimization and RDO vanishes. Other objective functions (like maximum stress) are suspect to show more distinct differences, but have not been implemented yet for topology optimization. Still, the example shows that first-order approaches may be used for robust topology optimization considering random material properties. For this example, the computational cost of the RDO is less than twice as much as the deterministic optimization due to the use of a first-order approximation.

Optimization approach	FOSM μ_c	FOSM σ_c	Monte Carlo μ_c	Monte Carlo σ_c
Deterministic	17.2	0.293	18.2	0.361
RTO with FOSM	17.0	0.283	18.0	0.349

Table 3.5. *Results of the cantilever beam example with random Young's modulus*

3.5. Conclusion and outlook

This chapter discusses first-order approximations at the mean vector of random input parameters, especially in the context of RDO. By application to different examples, it is demonstrated that:

– first-order approaches fail, if the objective function has a local minimum with respect to random parameters, which is present or can be reached during the optimization;

– whether design variables and uncertain parameters refer to the same physical property is alone not crucial for the applicability of first-order approximations;

– even if first-order approaches are not accurate, they may still be useful in RDO as they provide the correct tendency.

The first item, a local minimum w.r.t. random parameters, typically occurs for random load angle or fiber orientations of a composite ply. It also occurs for geometric deviations, if the perfect geometry is assumed to be the mean geometry. In contrast, there are parameters for which it is known that the objective function behaves monotonically. Then, first-order approaches work. This is the case for material properties like stiffness and strength parameters as well as wall thickness of shells or profile parameters of beams.

The main advantage of first-order approaches is the computational efficiency, when determining the required gradients with the adjoint method. Deriving these gradients for other, more sensitive objective functions is the subject of future research.

3.6. References

Arbocz, J. and Hol, J.M.A.M. (1991). Collapse of axially compressed cylindrical shells with random imperfections. *AIAA Journal*, 29(12), 2247–2256.

Asadpoure, A., Tootkaboni, M., Guest, J.K. (2011). Robust topology optimization of structures with uncertainties in stiffness – Application to truss structures. *Computers & Structures*, 89(11), 1131–1141 [Online]. Available at: http://www.sciencedirect.com/science/article/pii/S004579491000266X.

Balokas, G., Kriegesmann, B., Czichon, S., Böttcher, A., Rolfes, R. (2019). Metamodel-based uncertainty quantification for the mechanical behavior of braided composites. In *Advances in Predictive Models and Methodologies for Numerically Efficient Linear and Nonlinear Analysis of Composites*, Petrolo, M. (ed.). Springer, Cham.

Bendsøe, M.P. (1989). Optimal shape design as a material distribution problem. *Structural Optimization*, 1(4), 193–202 [Online]. Available at: https://link.springer.com/article/10.1007/BF01650949

Ben-Haim, Y. and Elishakoff, I. (1990). *Convex Models of Uncertainty in Applied Mechanics*. Elsevier, Amsterdam.

Beyer, H.-G. and Sendhoff, B. (2007). Robust optimization – A comprehensive survey. *Computer Methods in Applied Mechanics and Engineering*, 196(33), 3190–3218 [Online]. Available at: http://www.sciencedirect.com/science/article/pii/S0045782507001259.

Carrasco, M., Ivorra, B., Ramos, A.M. (2015). Stochastic topology design optimization for continuous elastic materials. *Computer Methods in Applied Mechanics and Engineering*, 289, 131–154 [Online]. Available at: http://www.sciencedirect.com/science/article/pii/S0045782515000444.

Cho, S.E. (2013). First-order reliability analysis of slope considering multiple failure modes. *Engineering Geology*, 154, 98–105 [Online]. Available at: https://linkinghub.elsevier.com/retrieve/pii/S0013795213000045.

Chryssanthopoulos, M.K. and Poggi, C. (1995). Probabilistic imperfection sensitivity analyses of axially compressed composite cylinders. *Engineering Structures*, 17(6), 398–406.

Chryssanthopoulos, M.K., Baker, M.J., Dowing, P.J. (1991). Imperfection modeling for buckling analysis of stiffened cylinders. *Journal of Structural Engineering*, 117(7), 1998–2017.

Cornell, C.A. (1969). A probability-based structural code. *Journal of the American Concrete Institute*, 66(12), 974–985 [Online]. Available at: https://www.concrete.org/publications/internationalconcreteabstractsportal/m/details/id/7446.

Dey, S., Mukhopadhyay, T., Adhikari, S. (2017). Metamodel based high-fidelity stochastic analysis of composite laminates: A concise review with critical comparative assessment. *Composite Structures*, 171, 227–250 [Online]. Available at: http://www.sciencedirect.com/science/article/pii/S0263822316328793.

Ditlevsen, O. and Madsen, H.O. (1996). *Structural Reliability Methods*. John Wiley & Sons, New York [Online]. Available at: http://od-website.dk/books/OD-HOM-StrucRelMeth-Ed2.3.7.pdf.

Dolinski, K. (1982). First-order second-moment approximation in reliability of structural systems: Critical review and alternative approach. *Structural Safety*, 1(3), 211–231 [Online]. Available at: http://www.sciencedirect.com/science/article/pii/0167473082900273.

Doltsinis, I. and Kang, Z. (2004). Robust design of structures using optimization methods. *Computer Methods in Applied Mechanics and Engineering*, 193(23), 2221–2237 [Online]. Available at: http://www.sciencedirect.com/science/article/pii/S0045782504000787.

Elishakoff, I. (1990). An idea of the uncertainty triangle. *The Shock and Vibration Digest*, 22, 1.

Elishakoff, I., van Manen, S., Vermeulen, P.G., Arbocz, J. (1987). First-order second-moment analysis of the buckling of shells with random imperfections. *AIAA Journal*, 25(8), 1113–1117.

Elishakoff, I., Kriegesmann, B., Rolfes, R., Hühne, C., Kling, A. (2012). Optimization and anti-optimization of buckling load for composite cylindrical shells under uncertainties. *AIAA Journal*, 50(7), 1513–1524.

Fragkos, K.B., Papoutsis-Kiachagias, E.M., Giannakoglou, K.C. (2019). pFOSM: An efficient algorithm for aerodynamic robust design based on continuous adjoint and matrix-vector products. *Computers & Fluids*, 181, 57–66 [Online]. Available at: http://www.sciencedirect.com/science/article/pii/S0045793018307680.

Fujita, K. and Takewaki, I. (2011). An efficient methodology for robustness evaluation by advanced interval analysis using updated second-order Taylor series expansion. *Engineering Structures*, 12, 3299–3310 [Online]. Available at: http://www.sciencedirect.com/science/article/pii/S0141029611003555.

Gournay, F.D., Allaire, G., Jouve, F. (2008). Shape and topology optimization of the robust compliance via the level set method. *ESAIM: Control, Optimisation and Calculus of Variations*, 14(1), 43–70 [Online]. Available at: https://www.cambridge.org/core/journals/esaim-control-optimisation-and-calculus-of-variations/article/shape-and-topology-optimization-of-the-robust-compliance-via-the-level-set-method/EC9F3ABB2BE02D3CB87D0D55773AB527.

Guest, J.K. and Igusa, T. (2008). Structural optimization under uncertain loads and nodal locations. *Computer Methods in Applied Mechanics and Engineering*, 198(1), 116–124 [Online]. Available at: http://www.sciencedirect.com/science/article/pii/S004578250800159X.

Guest, J.K., Prévost, J.H., Belytschko, T. (2004). Achieving minimum length scale in topology optimization using nodal design variables and projection functions. *International Journal for Numerical Methods in Engineering*, 61(2), 238–254 [Online]. Available at: https://onlinelibrary.wiley.com/doi/abs/10.1002/nme.1064.

Haldar, A. and Mahadevan, S. (1999). *Probability, Reliability and Statistical Methods in Engineering Design 1*. John Wiley & Sons, New York.

Hasofer, A.M. and Lind, N.C. (1974). Exact and invariant second-moment code format. *Journal of the Engineering Mechanics Division*, 100(1), 111–121.

Hong, Y.J., Xing, J., Wang, J.B. (1999). A second-order third-moment method for calculating the reliability of fatigue. *International Journal of Pressure Vessels and Piping*, 76(8), 567–570.

Jalalpour, M., Igusa, T., Guest, J.K. (2011). Optimal design of trusses with geometric imperfections: Accounting for global instability. *International Journal of Solids and Structures*, 48(21), 3011–3019 [Online]. Available at: http://www.sciencedirect.com/science/article/pii/S002076831100237X.

Kanno, Y. (2020). On three concepts in robust design optimization: Absolute robustness, relative robustness, and less variance. *Structural and Multidisciplinary Optimization* [Online]. Available at: https://doi.org/10.1007/s00158-020-02503-9.

Kriegesmann, B. (2012). Probabilistic design of thin-walled fiber composite structures. PhD thesis, Leibniz University Hannover, Germany [Online]. Available at: http://edok01.tib.uni-hannover.de/edoks/e01dh12/722293151.pdf.

Kriegesmann, B. (2020). Robust design optimization with design-dependent random input variables. *Structural and Multidisciplinary Optimization*, 61(2), 661–674 [Online]. Available at: https://doi.org/10.1007/s00158-019-02388-3.

Kriegesmann, B. and Lüdeker, J.K. (2019). Robust compliance topology optimization using the first-order second-moment method. *Structural and Multidisciplinary Optimization*, 60(1), 269–286 [Online]. Available at: https://doi.org/10.1007/s00158-019-02216-8.

Kriegesmann, B., Rolfes, R., Hühne, C., Teåmer, J., Arbocz, J. (2010). Probabilistic design of axially compressed composite cylinders with geometric and loading imperfections. *International Journal of Structural Stability and Dynamics*, 10(4), 623–644.

Kriegesmann, B., Rolfes, R., Hühne, C., Kling, A. (2011). Fast probabilistic design procedure for axially compressed composite cylinders. *Composites Structures*, 93, 3140–3149.

Kriegesmann, B., Rolfes, R., Jansen, E.L., Elishakoff, I., Hühne, C., Kling, A. (2012). Design optimization of composite cylindrical shells under uncertainty. *Computers, Materials, & Continua*, 32(3), 177–200 [Online]. Available at: http://www.techscience.com/cmc/v32n3/27911.

Lazarov, B.S., Schevenels, M., Sigmund, O. (2012). Topology optimization with geometric uncertainties by perturbation techniques. *International Journal for Numerical Methods in Engineering*, 90(11), 1321–1336 [Online]. Available at: http://onlinelibrary.wiley.com/doi/10.1002/nme.3361/abstract.

Lee, K.-H. and Park, G.-J. (2001). Robust optimization considering tolerances of design variables. *Computers & Structures*, 79(1), 77–86 [Online]. Available at: http://www.sciencedirect.com/science/article/pii/S0045794900001176.

Li, L. and Khandelwal, K. (2014). Two-point gradient-based MMA (TGMMA) algorithm for topology optimization. *Computers & Structures*, 131, 34–45 [Online]. Available at: http://www.sciencedirect.com/science/article/pii/S0045794913002812.

Luo, Y., Zhou, M., Wang, M.Y., Deng, Z. (2014). Reliability based topology optimization for continuum structures with local failure constraints. *Computers & Structures*, 143, 73–84.

Maute, K. and Frangopol, D.M. (2003). Reliability-based design of MEMS mechanisms by topology optimization. *Computers & Structures*, 81(8), 813–824 [Online]. Available at: http://www.sciencedirect.com/science/article/pii/S0045794903000087.

Metropolis, N. and Ulam, S. (1949). The monte carlo method. *Journal of the American Statistical Association*, 44(247), 335 [Online]. Available at: https://www.jstor.org/stable/2280232?origin=crossref.

Papoutsis-Kiachagias, E.M., Papadimitriou, D.I., Giannakoglou, K.C. (2012). Robust design in aerodynamics using third-order sensitivity analysis based on discrete adjoint: Application to quasi-1D flows. *International Journal for Numerical Methods in Fluids*, 69(3), 691–709 [Online]. Available at: https://onlinelibrary.wiley.com/doi/abs/10.1002/fld.2604.

Park, G.-J., Lee, T.-H., Lee, K.H., Hwang, K.-H. (2006). Robust design: An overview. *AIAA Journal*, 44(1), 181–191 [Online]. Available at: https://doi.org/10.2514/1.13639.

Schillo, C., Röstermundt, D., Krause, D. (2015). Experimental and numerical study on the influence of imperfections on the buckling load of unstiffened CFRP shells. *Composite Structures*, 131, 128–138 [Online]. Available at: http://www.sciencedirect.com/science/article/pii/S026382231500327X.

Schuëller, G.I. and Jensen, H.A. (2008). Computational methods in optimization considering uncertainties – An overview. *Computer Methods in Applied Mechanics and Engineering*, 198(1), 2–13 [Online]. Available at: http://www.sciencedirect.com/science/article/pii/S0045782508002028.

Schuëller, G.I. and Valdebenito, M. (2010). Reliability-based optimization – An overview. *Computational Technology Reviews*, 1, 121–155.

Sudret, B., Marelli, S., Wiart, J. (2017). Surrogate models for uncertainty quantification: An overview. *Proceedings of the 2017 11th European Conference on Antennas and Propagation (EUCAP)*, pp. 793–797.

Svanberg, K. (1987). The method of moving asymptotes – A new method for structural optimization. *International Journal for Numerical Methods in Engineering*, 24(2), 359–373 [Online]. Available at: http://onlinelibrary.wiley.com/doi/10.1002/nme.1620240207/abstract.

Wang, F., Lazarov, B.S., Sigmund, O. (2011a). On projection methods, convergence and robust formulations in topology optimization. *Structural and Multidisciplinary Optimization*, 43(6), 767–784 [Online]. Available at: https://link.springer.com/article/10.1007/s00158-010-0602-y.

Wang, X., Wang, L., Elishakoff, I., Qiu, Z. (2011b). Probability and convexity concepts are not antagonistic. *Acta Mechanica*, 219(1), 45–64. Available at: https://doi.org/10.1007/s00707-010-0440-4.

Wang, L., Liang, J., Wu, D. (2018). A non-probabilistic reliability-based topology optimization (NRBTO) method of continuum structures with convex uncertainties. *Structural and Multidisciplinary Optimization*, 58(6), 2601–2620 [Online]. Available at: https://doi.org/10.1007/s00158-018-2040-1.

4

Understanding Uncertainty

4.1. Introduction

Uncertainty is inherent in real life, whether it results from natural phenomena or from human action. It is a philosophical concept with which humankind is forced to live. After believing that feared events were the will of the gods to school or punish the mortals that we are, mankind learned to read this both scientifically and technologically. This chapter aims to explore the notion of uncertainty in confrontation that it imposes on designers and engineers in the definition of their products and systems during design and manufacture for optimal, robust, reliable and durable use.

We first look at the philosophical notion of uncertainty, then its relationship with design and we define the concept of a knowledge entity. Finally, we recall some approaches for robust and reliable design.

4.2. Uncertainty and uncertainties

"Everything that we can or should doubt is uncertain" writes André Comte-Sponville in his philosophical dictionary (Comte-Sponville 2001). "The fundamental uncertainty is that of our existence and it cannot be resolved since it would require the certainty of our reason". We live in uncertainty because we are confronted daily with questions about the events to come in the short or long term which can have several outcomes: Will it be sunny this afternoon? The uncertainty is a philosophical attitude that results from the existence of the numerous uncertainties that we are faced with every day: they surround us and are the multiple causes of our doubts, even our anxieties. Do such uncertainties result only from a lack of

---

Chapter written by Maurice LEMAIRE.

knowledge that could be resolved, therefore reducible, or are they intrinsic, therefore irreducible?

Reducible uncertainties are said to be epistemic (from the Greek word "*epistêmê*": knowledge). They relate to knowledge, i.e. our ability to limit the extent of the uncertainty through the accumulation of heuristic knowledge throughout the history of mankind or by the implementation of specific means of exploration, such as tests.

Irreducible uncertainties are often qualified as random or aleatory. Blaise Pascal tells us that the French "*hasard*", i.e. chance, has a structure. The geometry of chance is at the origin of the probability theory. This has led us to improperly associate randomness and probability because this theory is not the only possible representation of the uncertain universe.

The often-used distinction between epistemic and random (or aleatory) uncertainties has its origin in the physics of the infinitely small. The discussion between epistemic and random has been the subject of controversy between Albert Einstein and Niels Bohr. Does the quantum uncertainty result from an incomplete theory according to the former or from a fundamental random? The work of physicists today proves Niels Bohr right. Chance is due to quantum nature.

But what about in our fields of mechanics? For there to be a fundamental randomness, there would have to be a significant quantum of matter at the scale where we observe it, which is obviously not the case. Similarly, the attitude of a person is not, as some economists still believe, always reasoned to make the best decision based on his/her interest, but remains conditioned by education and learning, random sources of the space of possible decisions. In mechanics as in cognition, there is no intrinsic uncertainty, therefore irreducible, but a space in which the contribution of knowledge makes uncertainty reducible up to a level judged acceptable according to the stakes. What we call chance is only the gap between the knowledge, allowing us to obtain a forecast, and our observations; knowledge is the synthesis of our cognition and our uncertainties. There are no *a priori* epistemic or random uncertainties but uncertainties containing an epistemic and a random part. A representation model includes both parts with deepening levels depending on the consequences: deepen the epistemic in case serious consequences widen the randomness to increase generalization. Taleb (2007) invites us to consider the fractal nature of the sharing between cognition and uncertainty.

As Der Kiureghian and Ditlevsen (2009) point out, "the distinction between random and epistemic uncertainties is determined by the choice of our models". We could add, the choice of our tests. A test with a given protocol and given measuring instruments provides epistemic knowledge limited by the uncertainties of the protocol and the measurements.

In exploring uncertainty, it is impossible to start from the source and to follow all the Lagrangian trajectories ending in an observation domain at a given place and time (Figure 4.1), all the more so as they can be chaotic. It is only possible to make Eulerian observations at this place and at this time. In terms of uncertainties, we are necessarily Eulerian and our degree of latitude is the size of the field: wide, to take into account many situations or reduced with the risk of allowing no generalization.

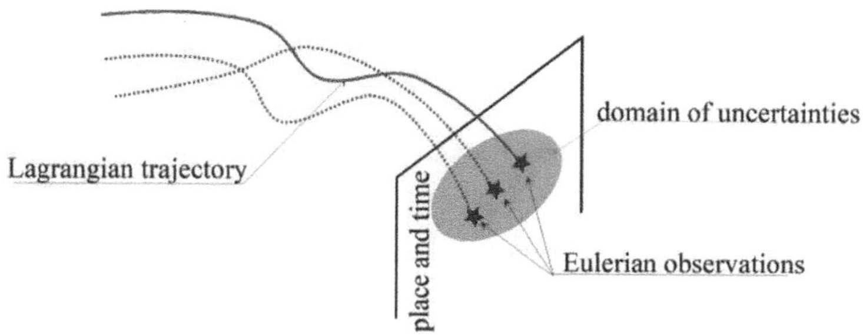

Figure 4.1. *Exploration of the uncertain domain at a place and time. For a color version of this figure, see www.iste.co.uk/challamel/mechanics3.zip*

Epistemic knowledge is marred by random uncertainty, particularly on the models and measurements, and randomness contains knowledge since Blaise Pascal wrote that chance has a geometry. There is therefore no need to distinguish between specific modeling in each case; however, a distinction must be made between what is reducible and what is not. When optimizing the mechanical design, it remains only a ranking of uncertainties according to the sometimes infinite expense of the effort to reduce them. Modeling must be tailored to knowledge of the uncertainty, both of the data and of the behaviors and be issue-driven in the decisions that the results must help to take. Some uncertainties are reducible by a contribution of new knowledge; others are not because the expense of investigation would be prohibitive or the ambition of the project would be too limited. There are therefore reducible and irreducible uncertainties.

4.3. Design and uncertainty

4.3.1. *Decision modules*

The question of uncertainty must be placed in the context of the life of a product or system, from the early stages of design, to those of project dimensioning and justification for robust and reliable operational functioning, and finally those of maintenance and decommissioning after use. Figure 4.2 shows the decision modules grouped into requirement, architecture, engineering and lifetime stages.

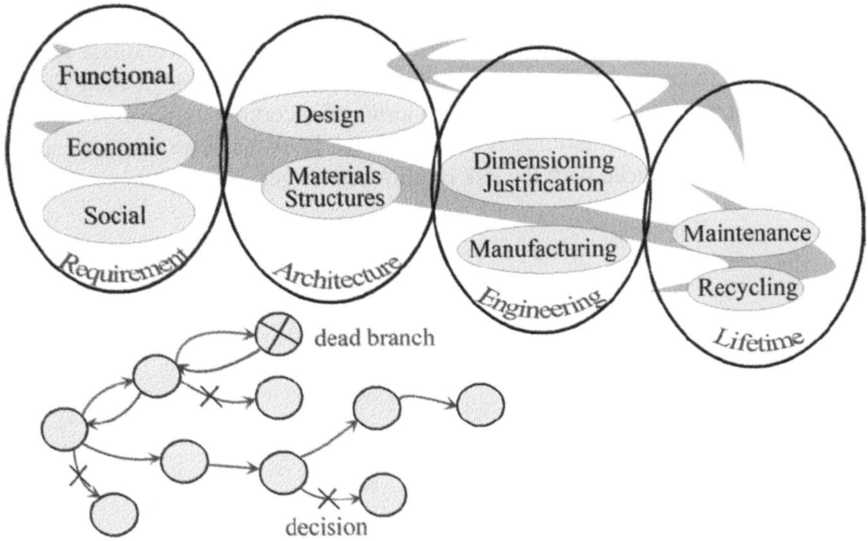

Figure 4.2. *Decision-making modules for creating an object and its life. For a color version of this figure, see www.iste.co.uk/challamel/mechanics3.zip*

The lifecycle stages proceed by anticipating the constraints of future stages by making only necessary and sufficient decisions at a given moment. However, it is obvious that the project progress cannot rely on certainties at each stage with the temptation of partial optimizations. The result is that every choice, every decision at any stage, must be taken among a set of possibilities constituting an uncertain design space. Such a space must be as wide as possible at each level so as to leave maximum freedom for future stages and thus allow maturation by the feedback loop, which will gradually restrict the possible choices under the constraints of cost, acceptability, robustness and reliability. A branch and bound diagram illustrates progression in the design cycle. A few initial implementations allow us to explore various branches (Figure 4.2), some of which can be deployed, others are eliminated by choice, and others are sterile and require backtracking.

4.3.1.1. *Requirements*

The starting point of a project is the definition of functional requirements, i.e. the specification of the objectives to be achieved. It should not be thought that, at this level, epistemic knowledge is sufficient. It is important to open up a space of possibilities so that the next steps can be carried out optimally. The wider this space is, the greater the possibilities for maturation and innovation. The knowledge entity (section 4.4) of the functional requirements is then a list of words or sentences, sometimes numerical values, and their satisfaction is a serial–parallel system among all possible systems.

It is contained in the preliminary design, for example building a permanent bridge to cross a gap.

The project must, of course, meet economic requirements of which the knowledge entity is the budget. Its space is uncertain, which is made up of numerical values associated with certainties (amount of the voted budgets) or uncertainties (expected subsidies, loans and expected revenues) as well as deadlines.

Finally, the project must have a societal requirement for its acceptability. The knowledge entity includes that which concerns nuisances, such as chemical or visual pollution, and also what enhances its context inclusion or sustainability.

The requirements are really uncertain variables since they are the result of choices from a set of possibilities, these choices are exercised either by the decision or by the maturation of the following stages of the design and life span.

4.3.1.2. *Architecture*

The architecture includes the design module which lists the possible solutions and a sketch of the technical choices in terms of materials and assembly, for example build a steel or concrete bridge.

The detailed design explores the uncertain space in order to estimate both technical and economic feasibility. It is an epistemic phase during which the project uncertainty is gradually reduced, while keeping as many possible open outcomes to address constraints motivated by the later stages of engineering justification and sustainability.

4.3.1.3. *Engineering*

The first module concerns the justification of the structural parameters under the imposed loads. It involves assessing the actions (maximum level and return period), the calculations according to more or less complex models and acceptance criteria (so-called safety coefficients), or even tests to validate innovations. This module is like those previously mentioned where the randomness is partly of the decision order and partly of the physics order, including an irreducible part. The second module is that of manufacturing where the knowledge entity identifies the possible solutions. Again, in the case of our example: loads, calculation code and construction method.

4.3.1.4. *Lifetime*

The decision modules concern the suitable lifetime of the product or system which involves performance, maintenance and dismantling or recycling at the service end. These modules, too often forgotten, neglected or deliberately sabotaged, as in the case of programmed obsolescence, act on the previous stages and are important requirements.

4.3.2. *Designing in uncertain*

Why was it necessary to extend uncertainty to the entire lifecycle, from the birth to the death of a product or system?

In the year 2020, during discussions with engineers at PHIMECA[1] on methods for assessing uncertainty in their products, they alerted us to the importance of introducing uncertainty management throughout the design chain, aiming for homogeneous levels of expertise. As the usual reasoning of many links is deterministic, it would then not be useful to occasionally introduce a stochastic approach; however, it is then useful to introduce such an approach in all the links, which is what is suggested here.

Figure 4.3 illustrates cognitive and modeling contexts for design under uncertainty. Cognitive science, i.e. the right, intuitive, imaginative, sensitive and global brain, is mobilized for acceptance. The hexagram in Figure 4.3, left, (Keyser *et al.* (1978)) depicts the relationships between the disciplines that allow us to move towards acceptance of the decision in uncertainty. Similarly, we propose (Lemaire (2014), Figure 4.3, right) a hexagram of the disciplines of the rational and sequential left brain that seeks to demonstrate this acceptance. As for any balanced being, design in uncertainty must mobilize both brains and their interactions: is a decision rational or emotional? The tools of thought are different and the exchanges are difficult. Modeling has a mathematical character associating a measure or a standard, whereas to say that an architectural choice is beautiful is to accept a realization according to an esthetic mode evolving in time and space.

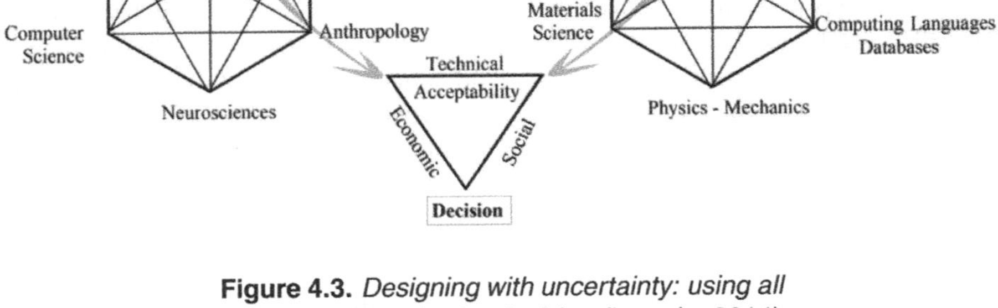

Figure 4.3. *Designing with uncertainty: using all resources to make the decision (Lemaire 2014)*

1 https://www.phimeca.com/en/

Making an approach to the uncertainty effective therefore requires an acculturation of all actors to the decision-making modules and their interactions. This implies defining knowledge entities whose modeling elements are mostly mathematical, as well as semantic. Thunnissen (2003) reminds us that reading uncertainty is very different from one discipline to another. However, the knowledge entities of cognition are beyond our engineering competence and will require dialogue with the proponents of the disciplines concerned, who will be able to guide us perfectly.

4.4. Knowledge entity

Defining a knowledge entity means defining a semantic object containing the available information. This definition must be as broad as possible and allow operations between entities. Today, we are satisfied by entities whose concepts are ultimately quite poor, even though it is a matter of probability theory. Indeed, probability theory is perfect when the space of uncertainty is known, as is the case in game theory where all the data are on the table, but not in engineering where all the distributions are conditional on the information available. Forgetting the uncertainty of the stochastic model would be a ludic fallacy according to Taleb (2007). A knowledge entity has a fractal granularity in accordance with the available information and the stakes of the decision to be taken. We are now interested in the engineering phase of the design cycle, for lack of being able to do better.

4.4.1. *Structure of a knowledge entity*

Item	Mapping
1	A list of words
2	A list of qualifiers
3	A list of possibilities
4	A Boolean statement
5	An expert value x_e
6	A characteristic or design value x_k or x_d
7	An interval $[x_{\min}, x_{\max}]$ or a convex
8	A quotation on a technical drawing
9	A probability distribution $X(\text{PDF}, m_X, \sigma_X, ...)$
10	A safety coefficient γ
11	An experimental measurement
12	...

Table 4.1. *Knowledge entity of the variable representation*

4.4.1.1. *Definition*

We call the knowledge entity of a variable (in a very broad sense), $X(x,t)$, a list of items that can gather the available information. It is indexed on time and space;

however these dimensions are not introduced here. Each item is appropriate to the use of the variable by a particular expert in a scientific or cognitive field. Table 4.1 is an example of the possible mapping of a variable by a few items. It is, of course, to be completed.

An entity has various levels as follows:

– It is said to be at level zero (basic entity $X^{(0)}$) if it corresponds to the level of lowest granularity from which the modeling is performed. The information is independent of other variables according to a physical relationship, but can be correlated in the statistical sense with other basic entities.

– It is higher than zero if it is the result of propagation from lower level variables:

$$X^{(l)}_{i=1,\ldots,n} \longrightarrow X^{(l+1)}$$

in which the passage from (l) to $(l+1)$ results from a combination of a mathematical algorithm, for example a stress $(l+1)$ is the quotient of a force (l) by a surface (l), and the cognitive algorithm that allows the architect to choose a design among all the possibilities identified in the first items.

4.4.1.2. *Entity level*

Let us note

$$X^{(l+1)}_k = \mathbf{M}\left(\underbrace{(\mathbf{F}, P_j)}_{\text{behavior model}}, \underbrace{X^{(l)}_{i=1,\ldots,n}}_{\text{data model}}\right)$$

where $\mathbf{M}$ denotes the operator of the combination model. It is a function of the entities reflecting the uncertainty of the data and of the choice of the representative base $\mathbf{F}$, itself a function of the knowledge entities P_j of the behavior model uncertainty. This base is expressed by various approaches: a simple analytical formula, an experimental formula, a polynomial, a neural network, a calculation code or simply, a list of words. In all cases, it includes uncertain coefficients P_j whose variability is adapted to $\mathbf{M}$'s validation domain. Thus, to establish the partial coefficients γ_j of a sizing rule, P_j must represent the variability over the entire application domain of the rule, whereas if it is a question of representing an experimental result, it contains only the measurement uncertainties at $X^{(l)}_{i=1,\ldots,n}$ fixed.

Building $\mathbf{M}$ is based on physical or cognitive knowledge in each profession and its calculation is the subject of a large amount of research work in the probabilistic context. $X^{(l)}_{i=1,\ldots,n}$ is known by a joint probability density; P_j is reduced to a given value and the density of $X^{(l+1)}_k$ is calculated, and then compared to a threshold to deduce a probability of exceeding it. What this chapter wants to underline is that this approach, which mobilizes strong mathematics, is insufficient to control the uncertainty. Attempts are made to widen the knowledge entity: fuzzy sets, possibility theory, probability boxes and so on, and this is how the methodology will be in accordance with the available information.

4.4.1.3. *Entity maturity*

According to the evolution of the design process, a knowledge entity goes through several levels depending on the stages of progress (Castric in Lemaire (2014)) as follows:

– The level of persistence p defines a scale from 1–5 of the degrees of validity of the information, i.e. the longevity of the information: not perennial, limited in time, valid for the duration of a study, valid over several programs and valid for current technologies.

– The level of variation v is represented on a scale of 0–3 of the quantification of the variation, i.e. the confidence that the information reaches its final value: very unstable, unstable, moderately unstable and stable.

– The sensitivity level s reflects the impact of the variability on the information on a scale from 0–3, described as: not sensitive, not very sensitive, moderately sensitive and sensitive.

– The completeness level c represents the combination of the depth (nature of the change) with the breadth of the information. Depth represents the nature of the change acting on the object (imprecision, abstraction and degree of detail). Breadth is the importance of the information in relation to the state of development expected by the user. Completeness represents the evolution of both dimensions. The four levels are as follows: incomplete, very partial, partial and complete.

The maturity of a knowledge entity is thus measured qualitatively by the quadruplet p, v, s, c. The elaboration and the follow-up of this quadruplet are a matter of know-how and being in adequacy with the stakes.

Table 4.2 is a very partial representation of the knowledge entity of a variable X. Each model considered – here, list, scalar, interval, possibilistic and probabilistic models – depends on a few parameters, to each of which is associated the quadruplet ($p = 1, ..., 5$ persistence, $v = 0, ..., 3$ variation, $s = 0, ..., 3$ sensitivity and $c = 0, ..., 3$ completeness) reflecting the quality of the information on each parameter. The maturity of the variable itself is a function $f(p, v, s, c)$. Thus, the maturity of a variable is expressed during the design process by a maturation of the information to reach a sufficient level:

$$X^{[i]} \rightarrow X^{[j]} \rightarrow X^{[k]} \rightarrow X^{[l]} \rightarrow X^{[m]} \rightarrow X^{[n]}, [i] < [j] < [k] < [l] < [m] < [n]$$

if an order relationship can be established. It will then be necessary to generalize this approach to variables that are a function of space and time. The limit towards which we must strive is an almost certain description, in which all the parameters of X are quoted ($p = 5$, $v = 3$, $s = 3$ and $c = 3$).

Variable	Model	Maturity	Parameters	Maturity
$X^{[i]}$	List	$[i] = f(p, v, s, c)$	Words	p, v, s, c
$X^{[j]}$	Scalar	$[j] = f(p, v, s, c)$	Value	p, v, s, c
	Safety coefficient		Value	p, v, s, c
$X^{[k]}$	Interval	$[k] = f(p, v, s, c)$	Min – max	p, v, s, c
$X^{[l]}$	Possibilistic	$[l] = f(p, v, s, c)$	Possibilities $\Pi(X)$	p, v, s, c
			Necessity $\mathrm{N}(X)$	p, v, s, c
$X^{[m]}$	Probabilistic	$[m] = f(p, v, s, c)$	PDF $F_X(x)$	p, v, s, c
			Mean m_X	p, v, s, c
			Confidence interval	p, v, s, c
			...	p, v, s, c
$X^{[n]}$	⋮	$[n] = f(p, v, s, c)$	⋮	⋮

Table 4.2. *Definition of a variable knowledge entity*

4.5. Robust and reliable engineering

4.5.1. *Definitions*

The knowledge entities of all design parameters and observation variables from their combination $X_{i=1,\ldots,n}^{(l),[k]}$ are the basis for demonstration of reliability and robustness.

Demonstrating reliability means showing that the expected operating point of the product or system in the design space is at a sufficient distance from potential points of failure. This involves first identifying all possible scenarios – this is the role of FMECA-type methods – and then measuring the distance in terms of number of standard deviations or reliability index. Demonstrating robustness means showing that around the target operating point, weak variability of the design variables leads to sufficient stability of the expected performance. These two criteria are linked to a degree of acceptability depending on the consequences of failure to achieve the objectives. We propose an illustration of the duality robustness–reliability. It is inspired by the famous example of Timoshenko in his book on elastic stability (Timoshenko 1961).

A ball subjected to gravity is placed in a parabolic-shaped cup (Figure 4.4). Its position is subject to a perturbation δx. If this perturbation is weak, the ball remains in the cup and returns naturally to its equilibrium position: the robustness is perfect. However, if the perturbation is large enough to make the ball come out of the cup, it is a failure scenario: the ball is ejected. There is a distance perturbation δx from which reliability is no longer assured. From a design point of view, the robust goal is that the ball stays in the cup and the reliable constraint is that it does not cross out the edges.

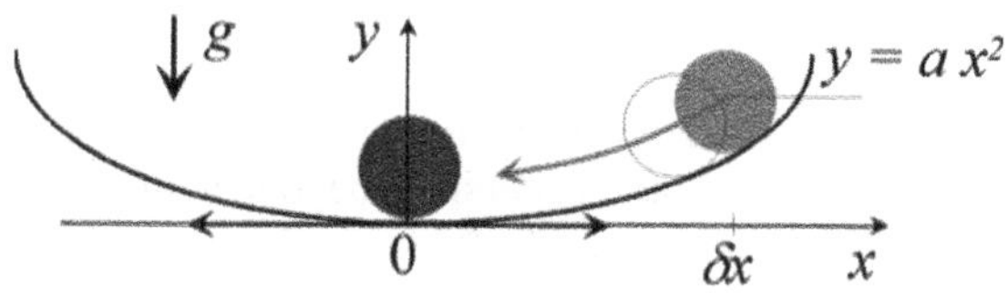

Figure 4.4. *An illustration of the concepts – robustness and reliability. For a color version of this figure, see www.iste.co.uk/challamel/mechanics3.zip*

Robustness is concerned with the ability of a system to adapt to small variations, whereas reliability is concerned with not reaching an unacceptable position.

4.5.2. Robustness

Let us consider a variable X that is susceptible to variation $2\Delta X$ and an objective function $Y(X)$ that expresses the response of a system whose minimum is sought. By placing ourselves at the optimal solution, the response is much more dispersed than by placing ourselves at the robust solution (Figure 4.5, left), where the objective function is almost flat. The curvature of the objective function is a good measure of robustness. To this data uncertainty, the model uncertainty must be added (Figure 4.5, right). The definition zone limited by physical bounds or probabilities invites us to place a robust solution where the zone is the tightest and the consequences on the response are the most limited. The illustration makes the robustness of the data and the model coincide; however, everyone can imagine other combinations. Above all, everyone must ask themselves, in the context of optimization, whether it is really carried out where the robustness is the greatest. It is generally sought to make a design choice robust, but not to place oneself where the design is the most robust.

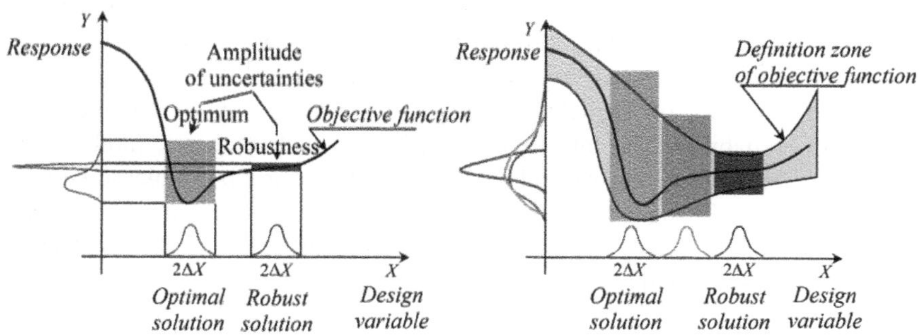

Figure 4.5. *Illustration of data robustness (left) and model robustness (right) (Lemaire 2014). For a color version of this figure, see www.iste.co.uk/challamel/mechanics3.zip*

4.5.3. *Reliability*

Robustness is concerned with the variability around the expected operating point. Reliability is concerned with potential excursions beyond a threshold between reliable situations and failure situations defined by $Y = f(X) > threshold$. From a data point of view (Figure 4.6, left), this amounts to underestimating the variable X_1 due to the negative derivative, while it amounts to overestimating X_2 due to the positive derivative. From the model point of view (Figure 4.6, right), reliability is ensured by shifting the objective function to the failure domain. This is traditionally done when experimental curves are shifted by two or three standard deviations.

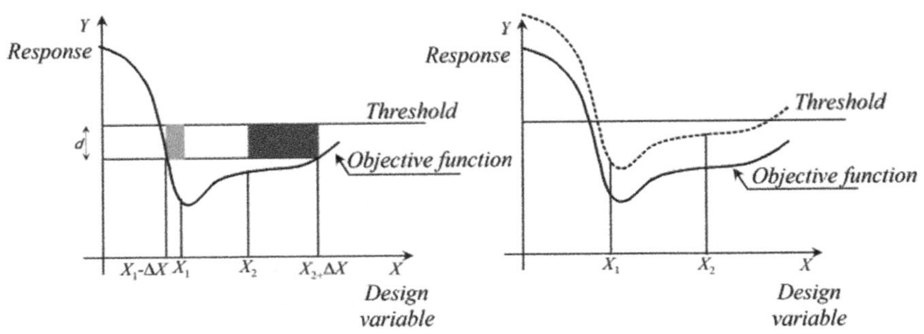

Figure 4.6. *Illustration of data reliability (left) and model reliability (right). For a color version of this figure, see www.iste.co.uk/challamel/mechanics3.zip*

4.5.4. *Optimization*

The first requirement of a design is to be reliable and robust according to the measures associated with these qualities. Reliability is an absolute constraint because it determines the risks to humans and the environment. Robustness is an objective to be satisfied as well as possible because it determines availability and maintenance. The pioneering work of Taguchi (1989) proposed a first answer to the duality robustness – reliability. The link between reliability and stochastic modeling was developed considerably in the second half of the 20th century; it is a pleasure to point out here that Isaac Elishakoff contributed to it as early as 1983, with a book whose third edition has recently been updated (Elishakoff 2017). His contribution highlighted here deals with optimization.

An optimization problem of a Ψ function reads simply:

$$X^{opt} = \arg\max \Psi(X_d); \;\; X_d \in D$$

and we must give meaning to the search space D according to the description of the knowledge entity X_d.

In Ben-Haim and Elishakoff (1990), Elishakoff used the interval theory and a convexity hypothesis: the design variables belong to an interval whose bounds are known, for example due to physical reasons. Within the convex domain, a worst-case principle selects an optimum domain by a procedure that Elishakoff and Ohsaki (2010) named "anti-optimization".

This approach is relevant when modeling of the knowledge entity belongs well to a certain interval; however, it does not exploit all the available stochastic knowledge, in particular the certainty of the bounds may lead us to consider large intervals and the convexity must be established. The procedure proposed in the following section classifies four optimization schemes with calculation variables extracted from the knowledge entity.

4.5.5. *Reliable and robust optimization*

According to the scheme initiated by Lelièvre *et al.* (2016), which associates robustness with a design objective $f(X)$ and reliability with a constraint $g(X)$, we can present the approaches (Figure 4.7).

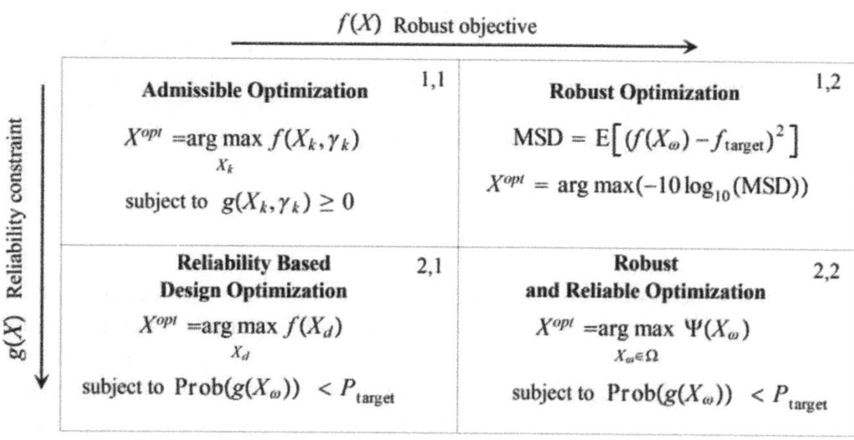

Figure 4.7. *Robust and reliable optimization*

First, it is necessary to extract computational variables from the knowledge entity: the design values X_d, the characteristic values X_k and the partial coefficients γ_k and in uncertain approaches, the variables X_ω to be sampled. There must exist a coherence between the deterministic model and the uncertain model which is only a higher level of knowledge maturation. The robust objective is generally a physical quantity that is easy to define, whereas reliability uses a probability measure with the great difficulty of defining whether it is acceptable or not. It is possible to accept a failure rate in the manufacture of a very large number of components, as long as the consequences of each failure are small; however, this is impossible, whatever the probability, if the consequences are catastrophic, in which case the constraint is to provide for the economic, human and environmental cost of a failure. However, we have only a notional probability objective, i.e. calculated with the available information.

Box (1,1) in Figure 4.7 is the classical problem for optimizing a design under constraint. This is how engineers proceed by using calculation codes inherited from experience, know-how and standards, enriched in recent years by calibration using probabilistic methods[2]. Box (1,2) gives the answer brought by Taguchi (1989) from experimental designs, generalized here in continuous formulation. Even though we must criticize the same weight between performance level and variability in the calculation of the MSD (Mean Square Deviation), Taguchi's relation brings a solution to robustness provided it is not sensitive to fragility, i.e. the target level f_{target} is well chosen. The (2,1) Reliability-Based Design Optimization (RBDO) box looks for a deterministic optimum for design values under a reliability constraint. Finally, Box (2,2) expresses the problem of robust optimization in uncertainty. The optimal solution maximizes an objective function Ψ built from the robustness requirements (Taguchi's or others) under the reliability constraint. This means that the optimum is computed for uncertain variables and meaning must be given to $X_\omega \in \Omega$. The problem is well posed for realizations of X_ω (Box 2,1) for each of which an optimum is obtained and a Pareto front should then make it possible to select the solution to be retained (Mercier *et al.* 2018).

4.6. Conclusion

Mastering uncertainty is an excellent field for the application of fine mathematical methods; however, it cannot be reduced to that. Through the contents of this chapter, we first wanted to remind ourselves that it requires many disciplines, other than those of the hard sciences, and that it means obtaining social, economic and technical acceptability of a decision involving risk of failure. The process of designing a product or system goes through stages of maturation, which enrich the entity of knowledge available on each parameter and their interactions. It remains to extract from this entity the variables allowing the forecasting calculations which are currently only possible with strong hypotheses. Finally, this design process must make it possible to demonstrate robustness and reliability, two complementary performances. It is in this field that a great deal of research is being carried out on the uncertainty quantification.

2 See: https://eurocodes.jrc.ec.europa.eu/

In *Association Française de Mécanique* (2015), on the future stakes of research in mechanics, a section was devoted by the author to scientific locks in mastering uncertainty; let us quote the following:

– the uncertain knowledge of behavioral models and their parameters;

– the modeling of uncertain parameters;

– the propagation of uncertainty in models related to decision-making;

– the global approach to multi-component, multi-scale and multi-physical performance;

and especially that of mind training and acculturation to the certainty of uncertainty, especially among young people in training.

This chapter, in the anniversary book dedicated to Professor Isaac Elishakoff, is an opportunity to thank him for our exchanges and for welcoming my students during internships and also, I hope, an opportunity to leave tracks for those who will continue our research.

4.7. References

Association Française de Mécanique (ed.) (2015). Mechanical engineering research – Industrial and societal issues: Research, innovation, training [in French]. White paper, EDP Sciences, Les Ulis.

Ben-Haim, Y. and Elishakoff, I. (1990). *Convex Models of Uncertainty in Applied Mechanics*. Elsevier, Amsterdam.

Comte-Sponville, A. (2001). *Philosophical Dictionary* [in French]. PUF, Paris.

Der Kiureghian, A. and Ditlevsen, O. (2009). Aleatory or epistemic? Does it matter? *Structural Safety*, 31, 105–112.

Elishakoff, I. (2017). *Probabilistic Methods in the Theory of Structures: Strength of Materials, Random Vibrations and Random Buckling*, 3rd edition. World Scientific, Singapore.

Elishakoff, I. and Ohsaki, M. (2010). *Optimization and Anti-Optimization of Structures Under Uncertainty*. Imperial College Press, London.

Eurocodes (2020). The EN Eurocodes [Online]. Available at: https://eurocodes.jrc.ec.europa.eu/.

Keyser, S.J., Miller, G.A., Walker, E. (1978). Cognitive science. Technical report, Alfred P. Sloan Foundation, New York.

Lelièvre, N., Beaurepaire, P., Mattrand, C., Gayton, N., Otsmane, A. (2016). On the consideration of uncertainty in design: Optimization – reliability – robustness. *Struct. Multidisc. Optim.*, 54, 1423–1437.

Lemaire, M. (2014). *Mechanics and Uncertainty*. ISTE Ltd, London, and Wiley, New York.

Mercier, Q., Poirion, F., Désidéri, J.-A. (2018). A stochastic multiple gradient descent algorithm. *European Journal of Operational Research*, 271(3), 808–817.

Taguchi, G. (1989). Introduction to quality engineering. Technical report, American Supplier Institute, Michigan.

Taleb, N.-N. (2007). *The Black Swan: The Impact of the Highly Improbable*. Random House, New York.

Thunnissen, D.-P. (2003). Uncertainty classification for the design and development of complex systems. *Proceedings from the 3rd Annual Predictive Methods Conference*. Newport Beach, California.

Timoshenko, S. (1961). *Theory of Elastic Stability*. McGraw-Hill, New York.

5

New Approach to the Reliability Verification of Aerospace Structures[1]

5.1. Introduction

An engineering design is a process of decision-making under constraints of uncertainty. This uncertainty is the result of a lack of deterministic knowledge about the different physical parameters and the uncertainty in the models that the design is performed with. Uncertainties exist in all engineering disciplines such as electronics, mechanics, aerodynamics, as well as structural design.

The uncertainty approach to the design of systems and subsystems was advanced by the engineering and scientific communities by the concept of reliability. Every system is now supposed to be analyzed for possible failure processes and failure criteria, probability of occurrence, reliability of basic components used, redundancy, possibilities of human errors in the production, as well as other uncertainties. The designer uses the contractual (or market) demands, with proper reliability appropriations for subsystems. The required reliability certainly influences both the design cost and the product cost.

Nevertheless, in most practical cases, structural analysts are still required to produce a structural design with absolute reliability, and most structural designs are performed using deterministic solutions, applying a factor of safety to cover for the uncertainties. The use of a safety factor is a *de facto* recognition in the random characters of many design parameters. Another approach is to use a worst-case analysis in order to determine the design parameters.

Chapter written by Giora MAYMON.

1 To Isaac – a friend and a colleague.

During the last three decades, the need for the application of probabilistic methods in structural design started to gain recognition and acceptance within the design communities. Structural designers started to use the stochastic approach and the concepts of structural reliability, thus incorporating the structural design into the whole system design. By the beginning of the 21st century, these approaches started to dominate many structural analysis procedures, incorporating the **structural design** into the total **system design**. In the last 30 years, theoretical and applied research has been developed to the extent that the probabilistic analysis of structures can use modern design tools (methods, procedures, computing codes) to apply the probabilistic approach to all practical industrial designs. The development of commercially available computer programs, which can be incorporated with the "traditional" tools, such as the finite elements computer codes, has contributed vastly to this practical progress. Dozens of textbooks (e.g. Lin 1967; Toft-Christensen 1982; Elishakoff 1983; Melcher 1999; Madsen *et al.* 2006) dealing with the probabilistic approach are now available, hundreds of new papers on the subject are published every year and conferences, which include non-deterministic approach sessions, are held. Thus, the field has reached a maturity, which justifies its routine use in the design process.

Methods to evaluate and assess the reliability of structural products were developed throughout the history of structural design. Some of these methods are old and do not include the new approaches established by the introduction of probabilistic approaches – probability of failure and stochastic safety factors. The purpose of this publication is to outline and suggest a new approach to evaluate the reliability of a structural design.

The reliability of a given design can be calculated using many modern design tools (e.g. NESSUS 2020, PROBAN 2020). Among them, the factor of safety and the computation of the probability of failure play an important role. These two terms are described in short in the next chapter, as a reminder of their importance.

5.2. Factor of safety and probability of failure

The factor of safety was introduced into the design process hundreds of years ago. An excellent historical survey can be found in Chapters 1 and 2 of Elishakoff (2004). It was introduced in order to compensate the designers for unknown design parameters or, in other words, uncertainties. Thus, although structural probabilistic analysis only started to gain access to design procedures in the last 40–50 years, the early designer used *de facto* a "device" to compensate for uncertainties in their knowledge of the models and the structures' properties, especially the allowable stresses.

This publication will not describe all of the aspects of the factor of safety. An extensive treatment of the subject, including history from ancient days, can be found in Elishakoff (2004). The reasons for the introduction of safety factors are discussed there.

It is commonly known that there are dispersions in many design parameters, both in the stresses (a function of the external loads, the geometry, the dimensions and the material properties) and the allowable stresses (a function of the material properties in the working environments). A nominal design with a small margin of safety may be unsafe due to the dispersion in many parameters, while a conservative design (with large safety factors) may be too heavy and sometimes more expensive.

One of the traditional methods to avoid failure in those cases where dispersions are expected is to do a "worst-case design". The values of the various variables are taken at the $\pm 3\sigma$ (σ is the standard deviation) of the dispersion range, and the structure is designed to survive these extreme values. Such an approach and its effect on simple design were described in Maymon (1998). An up-to-date approach to worst-case design is the use of a $\pm 6\sigma$ range for the random variables.

The use of a probabilistic approach in structural analysis is based on the assumption that the probability that **all** variables will be at their extreme values simultaneously is very low, its value depends on the variables' parameters and their inter-correlations. It is improbable that in a certain product all the extreme values will show up together. Therefore, it is the purpose of probabilistic structural analysis to predict the probability of failure of the structure, not to calculate its factor of safety. The failure criterion is determined by a probability of failure that should be smaller than a pre-defined value. The methods for applying probabilistic approaches to structural analysis were described in Maymon (1998, 2008). Development of methods of probabilistic structural analysis started in the 1970s, and the use of these methods gained popularity quite fast. Nevertheless, not all the aeronautical specifications and standards include such criteria, although many of them have been implemented into the civil engineering for many kinds of designs.

Usually, factors of safety are part of formal design codes and of the customer's specifications. Although the probabilistic approach to safety factors has many benefits over the classical approach, not many of the present aerospace formal design codes and specifications have adopted this approach yet. As design establishments are formally tied to the formal requirements, today, the probabilistic approach is only used in part of the design processes. Nevertheless, a combination of the classical safety factor approach with structural probabilistic methods can be adopted in order to both comply with the formal requirements and enjoy the benefits of the probabilistic safety factors analyses.

The classical and the probabilistic safety factor procedures can best be explained using the "stress–strength" model that was used extensively by many statisticians, and was well described in literature (e.g. Maymon 2008).

Suppose a structure is under a load S and has strength R. The load S is not necessarily the external load acting on the structure. S is understood as a required result of a structural parameter obtained in a structure of certain dimensions, material properties and external loads. It can be the stress in a critical location, the displacement obtained in the structure, the stress intensity factor due to crack propagation, the acceleration in a critical mount location, etc. The strength is not necessarily the material allowable. It can be the yield stress, the ultimate stress, the maximum allowed displacement due to contact problems, the fracture toughness of a structure, the buckling load of compressed member, etc. We can define a failure function

$$g(R,S) = R - S \qquad [5.1]$$

and define a failure when

$$g(R,S) \leq 0 \qquad [5.2]$$

Both R and S can be functions of other structural parameters; thus, the failure function, which describes a straight line in a 2D space, is generally a hyper-surface of n dimensions, which include all the variables that R and S depend on.

Using these notations, there are several possible definitions for the safety factor (e.g. Elishakoff 2004). The classical definition is

$$FS_{Classical} = \frac{R_{nominal}}{S_{nominal}} \qquad [5.3]$$

where the index "nominal" refers to given deterministic quantities.

When this number is larger than 1, there is no failure. The design codes demand a minimum value for $FS_{Classical}$ to avoid failure when uncertainties exist, say $FS_{Classical} \geq 1.2$.

Another possible definition for the safety factor is for a worst-case design:

$$FS_{WorstCase} = \frac{R_{minimum}}{S_{maximum}} \qquad [5.4]$$

In this case, the minimum possible value of R (which may be a function of several variables, or a value with known dispersion) is computed and divided into the maximum possible value of S. The values used here are the $\pm 3\sigma$ range. It is clear that $FS_{WorstCase}$ is smaller than $FS_{Classical}$. The advantage of $FS_{WorstCase}$ is that it takes the known dispersions in the structure's parameters into account. Its disadvantage is that it assumes that all the worst-case parameters exist simultaneously (although this may be a rare case, with a very low probability), and therefore, the resulted design is very conservative. Examples of the use of $FS_{Classical}$ and $FS_{WorstCase}$ were described in Maymon (2008), where it was demonstrated that two different design procedures can lead to inconsistent consequences, and also discussed the modifications required for a certain design when the worst-case safety factor is used.

Another possible definition for the safety factor is for a mean value design, sometimes called a central design:

$$FS_{Central} = \frac{E(R)}{E(S)} \quad [5.5]$$

E (.) is the expected value (the mean) of a parameter (.). In many cases, the mean values coincide with the nominal values, and the classical safety factor is identical to the central safety factor. For other cases where they do not coincide, the reader can explore the examples described in Elishakoff (2004).

All three of the safety factors defined in equations [5.3], [5.4] and [5.5] use deterministic values, and the safety factor obtained is deterministic.

The stochastic safety factor was introduced independently by Birger (1970) and in Maymon (2000, 2002), and in Elishakoff (2004), is called "the Birger–Maymon stochastic safety factor". The same concept was adopted later by others and is sometimes named the probabilistic sufficiency factor. The definition is

$$SF_{Stochastic} = \left(\frac{R}{S}\right) \quad [5.6]$$

As both R and S may be random variables (and can depend on many structural parameters that are assumed random), the obtained stochastic safety factor is also a random number. Using the stochastic safety factor, we may answer the question: "What is the probability that a structure (with given uncertainties) has a factor of safety smaller than a given value?"

The preceding definitions are demonstrated by a simple "stress–strength" example. Suppose R has a normal distribution with mean $\mu_R = 1.2$ and a standard deviation of $\sigma_R = 0.06$. S has a mean $\mu_S = 1.0$ and a standard deviation $\sigma_R = 0.05$. For both variables, the coefficient of variation (COV) is 5%. The classical safety factor is

$$FS_{Classical} = \frac{1.2}{1.0} = 1.2 \qquad [5.7]$$

which is an acceptable safety factor. For the worst-case design, assume that $\pm 3\sigma$ are the upper/lower limits of the variables. The worst-case safety factor is

$$FS_{WorstCase} = \frac{1.2 - 3 \cdot 0.06}{1 + 3 \cdot 0.05} = 0.887 \qquad [5.8]$$

which is unacceptable for the design.

The stochastic safety factor was computed using one of the probabilistic computer programs (some of them are listed in Maymon (2008, 2018). The results for the CDF (Cumulative Density Function) of this safety factor obtained using **first-order reliability methods** (known as FORM) are shown in Figure 5.1. The probability that the stochastic safety factor is equal to or smaller than 1.2 is 0.5. The probability that it is equal to or smaller than 1.0 is 0.0052226. It is interesting to see how the latter probability is changed if the distribution of the parameters is a truncated normal, where truncation is done at $\pm 3\sigma$ values. In this case, the probability that the stochastic safety factor is smaller than 1 is 0.0048687, a relatively small decrease in the probability of failure. Truncated normal distribution for a normally distributed variable means that a screening process is performed, and all specimens outside the $\pm 3\sigma$ range are screened out. The designer can do a cost-effectiveness analysis to determine whether the decrease obtained in the probability of failure is worth the much more expensive screening process.

When the coefficient of variation of both R and S is decreased, the effect of decreasing the dispersion in the random variables can be demonstrated. A decrease in the dispersion means tighter tolerances in the design and the production. In Figure 5.2, this effect demonstrates a significant decrease in the probability of failure.

In Maymon (2002, 2008), more sophisticated examples of factors of safety and calculations of probabilities of failure are described and treated.

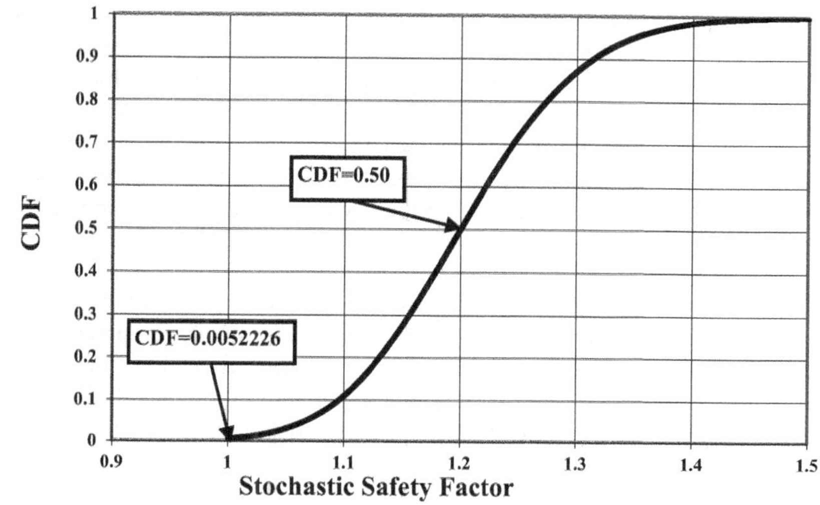

Figure 5.1. *CDF of the stochastic safety factor*

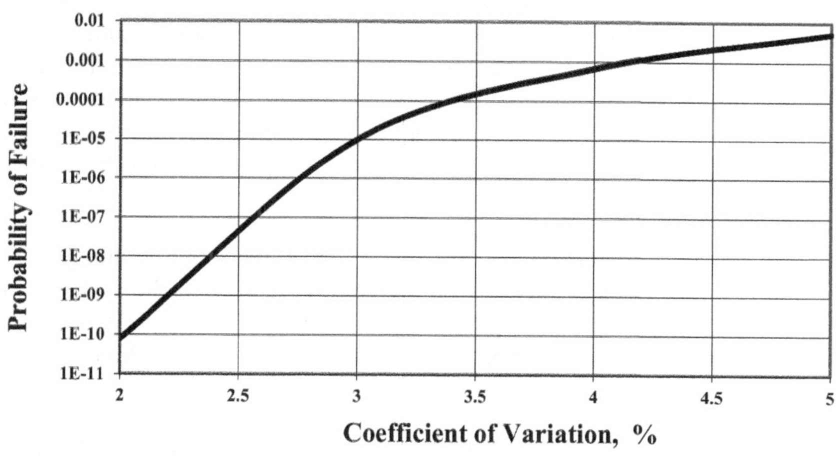

Figure 5.2. *Effect of COV on the probability of failure*

The importance of probabilistic analysis in the design process of structural elements is already well recognized. Analytical methods and computational algorithms were developed and are used in many design establishments and in R&D institutes. Randomness in structural geometric and dimensional parameters, material properties, allowable strength and external loads can now be treated during the design process. All of the methods described in the literature are based on a model (which may be analytical or approximated) or an algorithm, such as a finite element code. There may be questions about the validity of the model itself, which certainly has some uncertainties in it (e.g. Maymon 2005a). The model used in the solution of a problem does not always describe the behavior of the observed system, and in many cases, the discrepancy between the observed results and the model presents a random behavior. Well-known results are the buckling of a simply supported beam column or the propagation of cracks in metals (see Maymon 2018).

There is no way to avoid modeling in an intelligent design process. This is especially true for large projects, in which many subsystems comprise the final product, where the time to design and manufacture a prototype is long and when the number of tests is limited. Therefore, it is also important to formulate the uncertainty of the model, either by adding a suitable random variable or via a random process (e.g. Maymon 2005a, 2005b, 2012).

5.3. Reliability verification of aerospace structural systems

A structure is designed as a part of a whole system. In aircrafts, the system includes the disciplines of structures, aerodynamics, propulsion, control, electronics, chemical engineering, avionics, human engineering, production, maintenance and others. In missiles, explosive technologies are also added to include warhead design. During the design process of any system, a final reliability is required. This final reliability is obtained by using a combination of the reliabilities of all the components, sub-assemblies and subsystems, as well as production processes and methods.

The progress of the probabilistic analysis of structures achieved during the last three decades enables us to incorporate structural reliability in the system design process. The classical approach of the structural safety factor should be replaced with the reliability of the structure and its probability of failure. More on safety factors can be learned in the previous section.

There are some unique features that characterize aerospace systems (as well as large civil engineering projects) from consumer goods that clearly have to be reliable. The main difference is that these projects are of a large scale and of multidiscipline efforts. Another major difference is that in many cases, the final design is manufactured in a very small number of products (or systems). There are a very small number of space shuttles. Only one Hubble space telescope has been manufactured. Many satellites are "one of a kind" products. Only a small number of SR-71 intelligence aircrafts were manufactured. A relatively low number of ICBMs were produced. Only about seven dozen Cargo C-5 (Galaxy) aircrafts were originally manufactured. The designers of such projects cannot rely on statistical results obtained by testing many specimens, contrary to the designers and manufacturers of consumer goods. For the latter, statistical data can be obtained in both the design and the manufacturing phase, and a significant amount of experience feedback is obtained from consumers. As the number of final products is small, the development costs highly increase the unit price of an aerospace product, preventing (economically) the possibility of performing a large amounts of tests during this phase. Sometimes, the nature of the projects prevents testing in the final designed conditions, for instance, testing satellites in their space environment. In many cases, frequent modifications are introduced after the "end" of the development phase and two manufactured specimens can differ. Performance envelopes are usually very large and vary for different carriers. Usually, the cost of failure is high, in both performances and costs – and sometimes, in lives.

The true problem of these projects is not the computation of their reliability by mathematical tools, but the verification of the reliability values, sometimes called "reliability demonstration".

The demands for end-product reliability are traditionally expressed in a "required reliability" and a "required confidence level". It is common to find, in the customer's product specifications, a demand like "the required reliability is 90% with 90% confidence level". Although reliability engineers, statisticians and mathematicians may understand such a sentence, it is not clear to design engineers and project managers. For instance, it is well known to statisticians and reliability engineers that when **22 successful tests** of a system are performed, the 90% reliability with 90% confidence level is "demonstrated". It is less emphasized (and less commonly understood by project managers) that all 22 of these tests should be performed using the **same conditions**. When the project performance envelope is wide, several extreme "working points" should be tested (each with 22 successful tests), and this emphasizes the limitation of the classical demonstration process for such projects.

Today, the reliability of subsystems can be calculated ("predicted") by many techniques and methodologies. Then, the total reliability can be evaluated and predicted by combining the individual contributions of these subsystems and sub-assemblies, declaring that the predicted "reliability of 90% with 90% confidence level" is reached. Nevertheless, the real problem is not the **prediction** or the computation of the reliability by mathematical tools, but the **verification** of the reliability values, sometimes called "**reliability demonstration**". There is no way to "prove" or demonstrate that this reliability is really obtained. It is also hardly possible to convince customers, project managers and designers that the "confidence level" (a term they really do not understand) was also obtained.

It is well known that aerospace structures fail in service in spite of the **extensive (and expensive) reliability predictions and analyses on which much engineering effort is spent.** A well-known example is the failure of the Columbia space shuttle in 2003. Regretfully, two space shuttles failed in a total number of 100 flights. These failures set the shuttle flight project back many years, with a significant financial penalty, on top of the life and morale costs. Analysis of failures in many aerospace projects usually reveals that more than 80% of "field failures" are the result of a "bad" design that could have been avoided. Thus, improving the design process may significantly cut the amount of product failures, increase reliability, increase a product's safety and significantly decrease the costs. The main reasons for an erroneous design process are the use of inadequate design methodologies and a wrong design of tests and experiments during the development (design) process.

A different approach to the reliability demonstration of aerospace structures is required. Such an approach should be incorporated into the design process, by modeling the structural elements and structural systems and by performing tests to validate **the model**, and **not the product**.

The reliability of the final product should be deeply incorporated in the design process; thus, the **design engineers** have a significantly important role in the reliability demonstration process. The methodology of a highly improved design-to-reliability process must incorporate the expertise of the design engineers, together with the expertise of the statisticians and reliability engineers.

The "design-to-reliability" methodology suggested is based on the principles described and discussed in the following paragraphs.

5.3.1. *Reliability demonstration is integrated into the design process*

This principle implies a design team, which includes designers and reliability engineers, working together during the development phase. The reliability of the final product should be deeply incorporated in the design process; thus, the design engineers have a significantly important role in the reliability demonstration process, a role that is usually neglected today. The methodology of a highly improved design-to-reliability process must incorporate the expertise of design engineers together with that of statisticians and reliability engineers. The approach of "we designers will design and you, reliability engineers, will compute the reliability" should be discouraged. Thus, a methodology for a design process, which includes the reliability prediction and verification during the development phase is suggested.

5.3.2. *Analysis of failure mechanism and failure modes*

This is the most important phase in the design process, as it determines the main features of the design and failure criteria. The first analysis should be done during the conceptual design phase by the system engineers, with the participation of the designers, and updated during the full development phase. It is highly recommended that an additional independent failure analysis should be performed by experts who are not part of the project team, in order to use their experience. The project's system engineers and design engineers are the professionals who can best contribute to the process of the failure mechanism and failure mode analyses, not the reliability engineers. The latter can use their experience to contribute to the systematic process of failure mode analysis and direct the design engineers when performing this important design phase, but it is **the responsibility of the designers** to do the analysis. The failure mode analysis should direct the design process so that these failures will be outside the required performance envelope of the designed project. Even when failure mode analysis is done by teams of experts, clearly, some unpredicted failure modes will still remain, mainly due to lack of knowledge ("we did not think about it"). This implies that structural tests should be designed in configurations that would best simulate the mission profile of the system. This mission profile, whose preparation is also of major importance, should be prepared as early as possible in the project history and should be based on the project's specification and experience with similar products and projects gained in the past.

5.3.3. *Modeling the structural behavior, verifying the model by tests*

During the development tests, special experiments for the **model's verification** (rather than **product's verification**) are defined, designed and performed. In many cases, it is relatively simple to prepare a structural model (analytical or numerical) in which the structural behavior is examined and tests are performed. The model is then corrected and updated by the results obtained in these tests. In addition, the parameters that influence the structural behavior should be defined and verified by tests, data collection and the experience of both the design establishment and its designers. In cases where the structural model is not available, an empirical model can be built based on very carefully designed experiments that can check the influence of as many relevant parameters as possible. "Virtual tests" can then be performed, using the updated model to check the structural behavior in many points of the required working envelope. Results of these "virtual tests" can be included in the information required to establish the structural reliability.

5.3.4. *Design of structural development tests to surface failure modes*

Structural tests must be designed to surface one or more possible failure modes. The design of structural tests is an integrated and important part of the structural design process. It is hardly possible to perform one structural test that can simulate all the real conditions of the "project envelope" in the laboratory. Static loadings can be separated from dynamic (vibration) loadings, if the experimental facilities available cannot perform coupled tests, which is usually the case. It seems advantageous to perform coupled tests in which many failure modes can be surfaced simultaneously, but these kinds of tests are usually more complex and more expensive, and the results from one failure mode may obscure the outcomes of the other failure modes. Some of the structural tests that cause failure modes to surface may be common to the tests used for model verification, in order to decrease the costs of the structural specimens and shorten the structural testing time schedule. It is also recommended to experimentally establish a safety margin for the tested failure mode, as this margin can point out the extent of the structural reliability for the tested mode. Thus, structural tests should be **continued until failure** is obtained, unless a very high safety factor is demonstrated (when this happens, the design is not optimal). This approach may look too expensive and time-consuming to many project managers. In such cases, they should be encouraged to consider the price of failure at the more advanced stages of the development process, or after the product has already been supplied to the customer.

5.3.5. *Design of development tests to find unpredicted failure modes*

Tests should be conducted in "as real as possible" conditions. Thus, load locations, experimental boundary conditions and the tested structure should be designed as realistically as possible. The main difficulties may rise when vibration tests are performed on sub-assemblies and a complete structure. The need to introduce test fixtures contradicts the wish to perform realistic tests. Therefore, a new approach is called for the vibration tests methodology used presently and dictated by present specifications. This issue is a subject for quite a different discussion and will not be evaluated here. Some of the difficulties in performing realistic vibration tests were discussed in some detail in Maymon (2008).

5.3.6. *"Cleaning" failure mechanism and failure modes*

When failure in test occurs within the performance envelope (or outside it, but without the required safety factor), the design should be modified to "clean out" the relevant failure mode and the structure should be re-tested in order to verify the success of the "cleaning" process. The process of updating the design and its model must be repeated until no failure modes exist in the required performance envelope. This process may include "virtual tests" performed with the relevant model, verified by development tests. Such a process can assure, at the end of the development phase, that the reliability of the designed structure is very high, and qualitatively verified.

5.3.7. *Determination of required safety and confidence in models*

Safety margins or safety factors are defined at the beginning of a project and depend on its characteristics, the formal specifications and the past experience of the designers. It is recommended that in future projects, the approach of the stochastic safety factor (e.g. Elishakoff 2004; Maymon 2000, 2002) should be applied. Such an approach can bridge the gap between the classical safety factor used presently in most of the specifications and the probabilistic approach that started to gain recognition in the design community. The stochastic approach can provide a "translation language" between safety factors and reliability numbers. The confidence level is interpreted here as the confidence of the designer in the structural reliability obtained by the tests. This is not really a "mathematical statistical definition", but a design concept that has an engineering meaning, understood by customers, project managers and designers.

5.3.8. *Determination of the reliability by "orders of magnitude"*

The demonstrated reliability should be determined by "orders of magnitude" and not by "exact" numbers, while applying engineering considerations. The concept, which may be controversial, but has started to gain supporters in the design communities, is further discussed here.

In Figure 5.3, a flowchart of the design process is described using very general definitions. Of course, each of the blocks described in this chart can be further evaluated.

The demand for reliability demonstration defined by "order of magnitude" may be controversial, as it differs from the traditional "numerical demands" for reliability. Nevertheless, such a definition is much more realistic and provides a much better engineering-oriented approach to the issue of reliability demonstration. Traditionally, customers define, in their requirements, a numerical value for the product's reliability and for the reliability confidence level (i.e. reliability of 90% with 90% confidence level). Reliability engineers also use the same definition. For the large aerospace projects discussed, there is no meaning for such a requirement, as it is impossible to "demonstrate", verify or prove such values under the limiting circumstances of these projects. In addition, project managers and designers have some difficulties in translating the "confidence level" concept into a practical understanding of engineering.

Therefore, it is highly suggested to define reliability by "order of magnitude" such as "very high (A)", "high (B)", "medium (C)" and "low (D)". The "confidence level" concept can be replaced by the "designer confidence in the estimated reliability" (engineering confidence), such as "high confidence (a)", "medium confidence (b)" and "low confidence (c)". These definitions form a matrix, which is shown in Table 5.1.

The purpose of a high-reliability-oriented design is to "push" for the upper-left corner of the matrix. It can be argued that in order to "move" **upward** in the table, **design modifications** are required, while in order to "move" **leftward** in the table, **more tests** of the system and more evaluations of the model are required. There are no distinct sharp borders between the reliability levels and the confidence estimates in this matrix, as shown by the "undefined" lines in Table 5.1. In the last three columns of the table, numbers for traditional estimated reliability requirements are written. There is a difference in the demands from a sub-assembly, a subsystem (which comprises several sub-assemblies) and a system (several subsystems). The numbers are depicted in light gray to emphasize that they are not supposed to be "exact", and are only a required estimate for the required reliability. There is really no difference between a system with an estimated reliability of 99.9% and a system with an estimated reliability of 99.8%. On the other hand, it is certain that when the reliability of two different optional systems is examined, the first one showing a reliability of 99%, with the second one showing a reliability of 75%, the first system is more reliable than the second one. Exact numbers should therefore only be used as a qualitative comparison tool and not as an absolute quantitative tool.

The described approach is much more realistic than the classical one, which cannot be verified ("proved") for the kind of projects described. It puts much more emphasis on engineering considerations and concepts; therefore, it is much easier for a designer to understand and practice. The role of tests and experiments is major in the development process, and the importance of models and their verification becomes an important issue in the process.

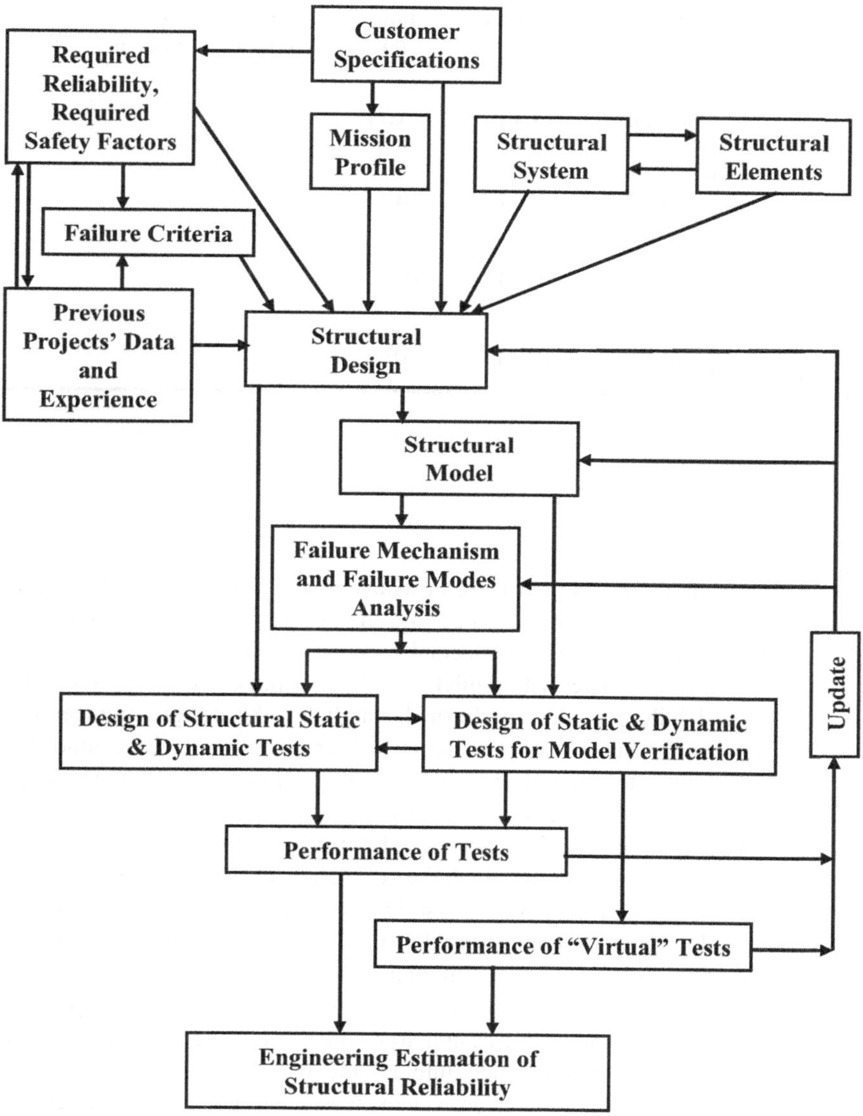

Figure 5.3. *Flowchart of structural design to reliability process*

The described (somehow controversial) approach was presented before an international audience in Maymon (2005b, 2012) and was well accepted by representatives **from the industry** in the audience. It was also applied successfully in the author's establishment.

Confidence Reliability	High (a)	Medium (b)	Low (c)	Sub- Assembly	Sub- System	System
Very High (A)	(A;a)	(A;b)	(A;c)	99.9	99	95
High (B)	(B;a)	(B;b)	(B;c)	99	95	90
Medium (C)	(C;a)	(C;b)	(C;c)	95	90	85
Low (D)	(D;a)	(D;b)	(D;c)	90	85	80

Table 5.1. *Reliability and engineering confidence matrix*

5.4. Summary

A design methodology is presented, in which structural reliability demonstration for large aerospace projects is integrated into the design process. The methodology is based on structural modeling, and special-purpose experiments were performed to verify and update the **structural models**, relevant parameters, analysis of failure mechanisms and failure modes (performed mainly by the design engineers). The design of structural tests whose purpose is to "surface" these failure modes, and tests, which may excite unpredicted failure modes, should also be done. Thus, a failure mode "cleanup" process is performed during the development tests. The determination of stochastic safety factors and the presentation of the demonstrated reliability by qualitative levels (in which the confidence of the engineer in his design is also expressed qualitatively) should also be performed. Some of the components of this methodology may be controversial, especially the qualitative analysis of the demonstrated reliability. Keeping in mind that the classical methodology really does not provide a realistic ("true") number for the demonstrated reliability, the presented approach can provide a powerful design tool, once a change in attitude towards demonstrated reliability is implanted into all the participants of the design and development process – customers, project managers, system engineers, reliability engineers and designers. At first glance, the suggested methodology may seem to be more costly than the traditional approach, mainly because of the demands for more tests (to validate the structural model) and for "tests to failure". Considering the significant cost of failures and the high cost of modifications required after such failures, the author believes that the total cost of the application of the described method will be much lower than the traditional approach.

The described approach, which was applied successfully in the author's establishment, can also be applied to other design disciplines (e.g. project safety demonstration) and should be treated in the system engineering levels, as it requires the cooperation of many groups in the design community. A change of attitude is required from management, customers, R&D establishments, industry and projects managers, in order to successfully apply the methodology to aerospace projects. The approach is **engineering-oriented**, and the academic and research communities are challenged to come forward with more rigorous tools for this engineering-oriented conceptual approach.

5.5. References

An inclusive list of hundreds of research works on the topic of structural reliability analysis could be prepared. Some of the major contributions are presented in this list of references.

Birger, I.A. (1970). Probability of failure, safety factor and diagnostics. *Problems of Mechanics of Solid Bodies*. "Sudostraenie" Publishers, Saint Petersburg.

Elishakoff, I. (1983). *Probabilistic Methods in the Theory of Structures*. John Wiley & Sons, New Jersey.

Elishakoff, I. (2004). *Safety Factors and Reliability: Friends or Foes*. Kluwer Academic Publishers, Amsterdam.

Lin, Y.K. (1967). *Probabilistic Theory of Structural Dynamics*. McGraw-Hill, New York.

Madsen, H.O., Krenks, S., Lind, N.C. (2006). *Methods of Structural Safety*. Dover Publications, New York.

Maymon, G. (1998). *Some Engineering Applications in Random Vibrations and Random Structures*. AIAA Inc., Progress in Astronautics and Aeronautics Series, 178.

Maymon, G. (2000). The stochastic safety factor – A bridge between deterministic and probabilistic structural analysis. *5th International Conference on Probabilistic Safety Assessment and Management*. PSAM -5, Osaka.

Maymon, G. (2002). The stochastic factor of safety – A different approach to structural design. *The 42nd Israel Annual Conference on Aerospace Sciences*, Tel-Aviv and Haifa.

Maymon, G. (2005a). What about uncertainties in the model? *46th AIAA/ASME/ASCE/AHS/ASC Structures, Structural Dynamics and Materials*, Austin, Texas.

Maymon, G. (2005b). Reliability demonstration of aerospace structures – A different approach. *Proceedings of the ICOSSAR 05 Conference*, Augusto, G., Schueller, G.I., Ciampoli, M. (eds). Rome.

Maymon, G. (2008). *Structural Dynamics and Probabilistic Analyses for Engineers*. B&H, Elsevier, Massachusetts.

Maymon, G. (2012). On the uncertainties of a structural model. *International Conference on Computational and Experimental Engineering and Sciences*, Crete.

Maymon, G. (2018). *Stochastic Crack Propagation – Essential Practical Aspects*. Academic Press, Elsevier, Massachusetts.

Melcher, R.E. (1999). *Structural Reliability Analysis and Predictions*. John Wiley & Sons Ltd, New Jersey.

NESSUS (2020). NESSUS-Structural probabilistic analysis code homepage. South West Research Institute, San Antonio, Texas [Online]. Available at: https://www.swri.org/nessus.

PROBAN (2020). The PROBAN structural probabilistic analysis code homepage. Det Norske Veritas, Oslo [Online]. Available at: https://manualzz.com/doc/7264297/proban---dnv-gl.

Thoft-Christensen, P. and Baker, M. (1982). *Structural Reliability and its Applications*. Springer-Verlag Berlin, Heidelberg.

6

A Review of Interval Field Approaches for Uncertainty Quantification in Numerical Models

6.1. Introduction

In mechanical engineering, current evolution in material and manufacturing technology opens up a vast world of possibilities. Combined with a state-of-the-art CAE-based virtual modeling environment, they enable us to design and build highly optimized products, meticulously engineered for optimal behavior and functionality, taking into account their broader system level, down to the material micro-scale. This evolution, however, brings with it many challenges for numerical design tools, as structurally complex (e.g. fiber-reinforced) or advanced manufactured (e.g. 3D-printed) products require material models with many parameters that are not always known with sufficient accuracy in the design phase. This uncertainty comes on top of more "classical" sources of uncertainty such as those present in the modeling of boundary conditions and structural loading conditions of the component. When considering a highly optimized design, accounting for these effects by introducing a large safety factor typically introduces conservatism, thereby impeding the optimization potential of the new technologies. Therefore, simulation tools that incorporate sources of uncertainty and variability, and thereby enable trustworthy judgment on design quality in the virtual design stage, are the core of a highly anticipated paradigm shift for optimized reliable design, and can even be considered a key enabler for the exploitation of modern material and manufacturing technology to their full potential.

Chapter written by Matthias FAES, Maurice IMHOLZ, Dirk VANDEPITTE and David MOENS.

Non-deterministic approaches that enable uncertainty analysis in numerical simulation have been studied extensively over the past decades. In engineering practice, so far, these tools are mainly used to assess variability between different realizations of a product (= inter-variability). However, taking into account the current trends in materials and manufacturing technology as described above, new challenges are surfacing related to the spatial nature of the uncertainty within the realization of the product (= intra-variability). The description of property variations within the product geometry, however, requires an approach that differs strongly from the parametric studies mentioned above. On the one hand, the required user interaction reaches far beyond the current practice of defining distributions or bounds on individual model parameters, as the spatial dependency of the variability needs to be quantified as well. On the other hand, current computational capabilities drive designers to use ever more detailed numerical models, with mesh sizes ranging up to millions of degrees of freedom. In order to enable an equally detailed spatial description of the uncertain model properties, these have to be defined in a correspondingly high-dimensional context. As this typically results in a large number of non-deterministic variables, this also complicates the non-deterministic analysis phase.

Typically, non-deterministic models of spatial uncertainty are modeled in the well-established random field framework (Vanmarcke and Grigoriu 1983). However, while tailored for exactly this type of spatial variation, the framework currently has only limited success in industrial engineering practice. This is mainly caused by its computational burden, which renders the analysis of industrially-sized problems very challenging, even when resorting to highly efficient random field analysis methods such as the expansion optimal linear estimator (EOLE) method (Liu and Der Kiureghian 1991). Aside from that, the methodological complexity, high information demand and rather indirect control of the spatial variation have also limited its cost–benefit potential for prospective end-users. Especially this last point, which is related to the fact that a random field usually relies on a global, rather than local, correlation metric (due to the widely applied stationarity assumptions), renders the method less appropriate for modeling manufacturing effects, as these will typically influence the material very locally. The data requirement aspect was recently relaxed by the introduction of imprecise random fields (see Dannert *et al.* (2018) or Faes and Moens (2019a)), which allow us to explicitly account for uncertainty in the definition of the random field model.

This chapter focuses on the use of the more intuitive interval concept in the context of modeling spatially uncertain properties and complements a recent review paper by the authors (Faes and Moens 2019a). As such, it presents an overview of the current state-of-the-art, including a discussion on the available interval field concepts, the different approaches for their quantification in practice, as well as the propagation aspects, discussed from an engineering perspective. Section 6.2 discusses the basics of interval finite element analysis. Section 6.3 lists recent developments in the context of set-based finite element analysis. Section 6.4 discusses the state-of-the-art in interval field finite element analysis. Finally, section 6.5 lists the conclusions.

6.2. Interval finite element analysis

Let $\mathcal{M}$ be a deterministic numerical model that is used to approximate $\boldsymbol{y} \in \mathbb{R}^{n_y}$, the solution of a (set of) differential equations, through a set of (usually) real-valued function operators $\boldsymbol{g} = \{g_i \mid i = 1, \ldots, n_y\}$:

$$\mathcal{M}(\boldsymbol{x}) : y_i = g_i(\boldsymbol{x}), \qquad g_i : \mathbb{R}^{n_x} \mapsto \mathbb{R}, i = 1, \ldots, n_y \tag{6.1}$$

with $\boldsymbol{x} \in \mathcal{F} \subset \mathbb{R}^{n_x}$ the vector of model parameters and $\mathcal{F}$ the sub-domain of feasible parameters. Following an interval approach, the uncertainty that is attributed to $\boldsymbol{x}$ is modeled as an interval vector $\boldsymbol{x^I} \in \mathcal{F}^I \subset \mathbb{IR}^{n_x} = x_1^I \times x_2^I \times \ldots \times x_{n_x}^I$, with $\mathbb{IR}^{n_x}$ the space of n_x-dimensional interval vectors and $\times$ denoting the Cartesian product. Therefore, by construction, $\boldsymbol{x^I}$ spans a hyper-rectangular space in $\mathbb{R}^{n_x}$ and hence, all x_i, $i = 1, \ldots, n_x$ are orthogonal, and, as such, independent. The interval FE method generally aims at finding a solution set $\tilde{\boldsymbol{y}}$ containing the extreme realizations of $\boldsymbol{y^I}$, resulting from propagating $\boldsymbol{x^I}$, which is defined as:

$$\tilde{\boldsymbol{y}} = \{\boldsymbol{y} \mid \boldsymbol{x} \in \boldsymbol{x^I}, \boldsymbol{y} = \mathcal{M}(\boldsymbol{x})\} \tag{6.2}$$

In general, $\tilde{\boldsymbol{y}}$ spans a non-convex manifold in $\mathbb{R}^n_y$, since $\mathcal{M}$ provides possibly nonlinear coupling between at least a subset of y_i, $i = 1, ..., n_y$. In general, finding an exact representation for $\tilde{\boldsymbol{y}}$ is an NP-hard problem. Therefore, a considerable amount of research is dedicated toward solving the interval propagation problem. Mostly, methods are aimed at finding an interval vector $\boldsymbol{y^I} = [\underline{\boldsymbol{y}}, \overline{\boldsymbol{y}}] \supseteq \tilde{\boldsymbol{y}}$, where each $\underline{y}_i$ and $\overline{y}_i$ represent the lower and upper bounds of each separate response. Following the anti-optimization framework (Qiu and Elishakoff 1998), these bounds are obtained as:

$$\begin{aligned} \underline{y}_i &= \min_{\boldsymbol{x} \in \boldsymbol{x^I}} g_i(\boldsymbol{x}) \qquad i = 1, ..., n_y \\ \overline{y}_i &= \max_{\boldsymbol{x} \in \boldsymbol{x^I}} g_i(\boldsymbol{x}) \qquad i = 1, ..., n_y \end{aligned} \tag{6.3}$$

The function operators $\boldsymbol{g} = \{g_i \mid i = 1, \ldots, n_y\}$ can either be provided directly by the numerical model under consideration or represented by surrogate models such as Kriging (see, for example, De Munck *et al.* (2009) or Khodaparast *et al.* (2011)), artificial neural networks (see, for example, Broggi *et al.* (2017), Bogaerts *et al.* (2019) or Faes *et al.* (2019a)), interval predictor models (see Crespo *et al.* (2014) for a theoretical explanation and Faes *et al.* (2019b) and Sadeghi *et al.* (2020) for applications in uncertainty quantification), polynomial response surfaces (Sofi and Romeo 2018), Chebyshev-based series expansions (see Wu *et al.* (2013) or Li *et al.* (2017)), Taylor series expansions (Fujita and Takewaki 2011) or a dimension-wise approach (Wang *et al.* 2018). For monotonic problems, the problem is often solved using a combinatorial approach, also known as the transformation method for fuzzy uncertainty (see Hanss (2005)), as, in this case, it yields the exact solution. The key idea here is to interpolate the uncertainty linearly between all combinations of vertices of the input hyper-rectangle $\boldsymbol{x}^I$. This method was also recently extended to the propagation of multivariate, dependent interval uncertainty in Faes and Moens (2019b). The main drawback of this method, however, lies in the combinatorial scaling of the computational cost of the propagation.

An alternative approach lies in the direct solution of equation [6.3] by replacing the *regular* algebraic expressions that constitute the solution sequence of $\mathcal{M}(\boldsymbol{x})$ by *interval algebraic* expressions. Specifically, in the case of an interval finite element model, the interval-valued parameters $\boldsymbol{x}^I$ are translated to interval element stiffness $\boldsymbol{K}_e^I$ and mass element matrices $\boldsymbol{M}_e^I$ according to general finite element formulations (in, for example, an undamped dynamical model). Then, these interval element matrices are assembled into the interval system matrices, $\boldsymbol{K}^I$ and $\boldsymbol{M}^I$ for, respectively, the stiffness and mass matrices and these interval system matrices are used to approximate the solution of the analysis. The main advantage of the technique is its numerical efficiency because, as opposed to global optimization approaches, no iterative sampling of the deterministic numerical model is needed for the solution of the problem. However, an overestimation of the *true* interval width occurs in general since intervals cannot track parameter dependencies by definition. This overestimation originates from multiple occurrences of the same interval parameter in the arithmetic operations, and stems directly from the assumption that interval numbers are independent. This is also referred to as the *dependency phenomenon* (Muhanna and Mullen 2001). In practice, the degree of overestimation is proportional to the width of the intervals x^I and y^I and the number of uncertain parameters (Sofi and Romeo 2016).

A solution to the dependency phenomena lies in the application of *affine arithmetic* instead of regular interval arithmetic, as introduced by Manson (2005). The principal idea of this method is to represent all n_x interval parameters x using their affine form:

$$\langle x^I \rangle = x_0 + \sum_{i=1}^{n_x} x_i \hat{\epsilon}_i^I + x_e \hat{\epsilon}_e^I \qquad [6.4]$$

with $\hat{\epsilon}_i^I \in [-1, +1]$ the unknown symbolic real independent interval variables, which allow for keeping track of dependency through addition, subtraction and scalar multiplication. $x_e \hat{\epsilon}_e^I$ is an error term introduced to account for possible nonlinear dependencies (Degrauwe *et al.* 2010). Muscolino and Sofi (2012) further extended the ideas of Manson and elaborated on the symbolic interval variable, which they denote as the *extra unitary interval* (EUI). It is defined such that the following properties hold:

$$\hat{\epsilon}_i^I - \hat{\epsilon}_i^I = 0 \qquad [6.5a]$$

$$\hat{\epsilon}_i^I \times \hat{\epsilon}_i^I \equiv (\hat{\epsilon}_i^I)^2 = [0, 1] \qquad [6.5b]$$

$$\hat{\epsilon}_i^I \times \hat{\epsilon}_j^I = [-1, +1] \quad i \neq j \qquad [6.5c]$$

$$x_i\hat{\epsilon}_i^I \pm y_i\hat{\epsilon}_i^I = (x_i \pm y_i)\hat{\epsilon}_i^I \qquad [6.5d]$$

$$x_i\hat{\epsilon}_i^I \times y_i\hat{\epsilon}_i^I = x_i y_i(\hat{\epsilon}_i^I)^2 = x_i y_i[0, 1] \qquad [6.5e]$$

with x_i and y_i finite numbers associated with the ith EUI, $\hat{\epsilon}_i^I$. By associating an EUI to each interval variable, the dependency can be taken into account through the computations. Moreover, the over-conservatism in the matrix assembly phase is alleviated as the interval radius of the stiffness $\Delta \boldsymbol{K}$ can be written as the superposition of the contribution of each separate interval parameter x_i^I (see Sofi and Romeo (2016) for the proof). The subsequent finite element analysis can then be performed by means of an interval rational series expansion (Sofi and Romeo 2016). The applicability of the *improved interval analysis via extra unitary interval* has been demonstrated in the context of interval perturbation in Muscolino and Sofi (2012), interval arithmetic computations of truss structures in Muscolino and Sofi (2013) or for the computation of natural frequencies of structures containing interval non-determinism in Sofi *et al.* (2015a).

6.3. Convex-set analysis

The independence between multiple intervals can also yield over-conservative estimates of the bounds of a structural response in cases where multiple interval-valued parameters are coupled due to physical phenomena. For instance, considering parameters such as material strength and stiffness of a component that is produced using a casting approach, typically a positive dependence between such parameters would be expected. Conversely, when looking at the width and thickness of such a part, a negative dependence could be introduced due to gravitational effects.

An intuitive approach to deal with such cases of uncertainty modeling is based on the set-theoretical work of Elishakoff and his colleagues, who throughout recent years introduced several set-theoretical approaches to cope with dependence in a non-probabilistic way (Zhu *et al.* 1996; Wang *et al.* 2008). Following the most basic approach, the admissible set $\mathcal{D}$ of two jointly dependent interval variables can be represented using a d-dimensional hyper-ellipsoid, which should abide by some minimum volume property (Elishakoff and Archaud 2013). Also, extensions towards

Lamé curves and other, nodal, convex sets were introduced in recent years (Elishakoff and Sarlin 2016a, b). Such set-based descriptors have also proven their merit in the inverse identification of multivariate interval uncertainty based on small sets of indirect measurements (Faes and Moens 2017; Faes *et al.* 2017).

However, while providing the analyst with an intuitive tool, the underlying assumption is still that all parameters are governed by a single underlying dependence structure. Recent work of the authors expanded "classical" set-theoretical methods to allow for defining intricate dependence structures in an intuitive way. To do this, techniques that are commonly applied in the context of constructing copula pair constructions are translated to an interval context (Faes and Moens 2019b). Further recent continuations of the seminal work of Elishakoff include applications in reliability-based design optimization (Li *et al.* 2019a), hybrid seismic risk analysis (Liu and Elishakoff 2019), hybrid analysis (Zhan *et al.* 2020) or buckling analysis (Verhaeghe *et al.* 2013b).

6.4. Interval field analysis

Consider in this context a model input quantity $x(\mathbf{r})$ that varies over a discretized d_d-dimensional domain $\Omega \subset \mathbb{R}^{d_d}$, indexed by a parameter $\mathbf{r} \in \Omega$. This model domain can be either space, time or a combination of the two. To propagate this quantity following the classical interval framework, two approaches are usually applicable. A first approach attributes a single interval to each discrete element $\Omega_e \subset \Omega, e = 1, ..., N_e$, with Ω_e for instance corresponding to the N_e finite elements of the finite element model $\mathcal{M}(\mathbf{r})$ or time instances of a time series. This, however, provides the analyst with spurious results, as unrealistic realizations are explicitly taken into account in the analysis, since intervals are independent by nature, and hence, realizations of the uncertain field with large spatial discontinuities are explicitly included in the analysis. Furthermore, such an approach inflates the dimension of the hyper-rectangular space drastically, which makes application to even medium-sized models intractable due to the so-called *curse of dimensionality*. On the other hand, an analyst can attribute a single interval $[\min_\Omega (x(\mathbf{r})) ; \max_\Omega (x(\mathbf{r}))]$ that captures the extremes over Ω fully. While very efficient, this neglect of the varying nature of the quantity might yield severely over-conservative results, especially when the uncertain model quantity is not homogeneously uncertain over the model domain (see, for example, Faes and Moens (2017)). Spatial interval uncertainty has also been discussed in Wu and Gao (2017). In recent work, Moens *et al.* (2011) introduced the concept of *interval fields* as an interval counterpart to the established framework of random fields. Interval fields counteract these problems by imposing an auto-dependence structure on the spatio-temporal distribution of the uncertain property under consideration. Since then, several techniques for modeling spatio-temporal uncertain model quantities in an interval context have been introduced. Three groups of techniques can be identified: the explicit interval field formulation, interval fields based on the Karhunen–Loève (KL) expansion and interval fields based on convex descriptors. The following sections will elaborate on these methods in detail.

An example of the bounds of an interval field is shown in Figure 6.1. This figure shows an upper and lower bound of the spatial uncertainty on the thickness t of a suspension arm, produced via additive manufacturing (for more details, the reader is referred to Faes *et al.* (2019c)). Within these bounds, t can vary freely, while taking into account the dependence imposed by the interval field.

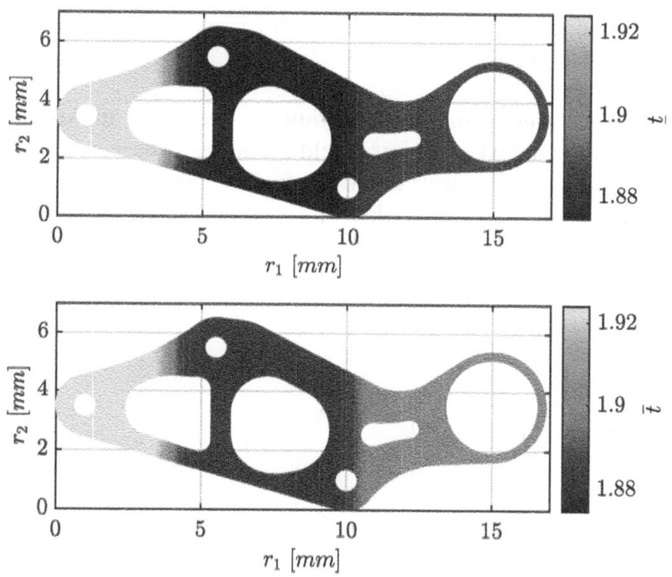

Figure 6.1. *Upper and lower bounds of spatial uncertainty on the thickness t of a suspension arm, produced via additive manufacturing (taken from Faes* et al. *(2019c)). For a color version of this figure, see www.iste.co.uk/challamel/mechanics3.zip*

6.4.1. *Explicit interval field formulation*

Following the explicit formulation, an interval field $\boldsymbol{x}^I(\mathbf{r}) : \Omega \times \mathbb{IR}^{n_b} \mapsto \mathbb{IR}$ is defined as:

$$\boldsymbol{x}^I(\mathbf{r}) = \sum_{i=1}^{n_b} \psi_i(\mathbf{r})\alpha_i^I \qquad [6.6]$$

and can be interpreted as the superposition of $i = 1, \ldots, n_b \in \mathbb{N}$ base functions $\psi_i(\mathbf{r}) : \Omega \mapsto [0, 1]$, which represent a set of spatial uncertainty *patterns*, scaled by independent interval scalars $\alpha_i^I \in \mathbb{IR}$, which represent the magnitude of the uncertainty. For a more in-depth explanation, the reader is referred to the work of Verhaeghe *et al.* (2013a).

In practical applications of the interval field framework, the base functions $\psi_i(\mathbf{r})$ in [6.6] should translate expert knowledge of the analyst on the varying nature of the uncertainty to a mathematical formulation in an intuitive way, while delivering a realistic representation of this uncertainty. The definition as such can be based on engineering judgment or on direct (see Imholz *et al.* (2018)) or indirect (see Faes and Moens (2017)) measurement data. Two methods for the construction of $\psi_i(\mathbf{r})$ were introduced very recently by the authors: inverse distance weighting interpolation and local interval field decomposition.

Inverse distance weighting interpolation was recently applied by Faes and Moens (2017) in the context of interval field modeling. The core idea is to define intervals α_i^I at appropriate control point locations $\mathbf{r}_i$ inside the model domain Ω. Then, the base functions $\psi_i(\mathbf{r})$ interpolate this local uncertainty toward the rest of the model domain, based on the inverse of the distance to each of these control points. Specifically, the base functions $\psi_i(\mathbf{r})$ are constructed as:

$$\psi_i(\mathbf{r}) = \frac{w_i(\mathbf{r})}{\sum_{j=1}^{n_b} w_j(\mathbf{r})} \qquad [6.7]$$

with $w_i(\mathbf{r}) \in \Omega$ and $i = 1, \ldots, n_b$:

$$w_i(\mathbf{r}) = \frac{1}{[d(\mathbf{r}_i, \mathbf{r})]^p} \qquad [6.8]$$

with $p \in \mathbb{R}^+$. Based on this approach, the number of control points and their location in Ω directly affects the spatio-temporal nature of the interval field realizations. As such, these parameters can be either tuned by an analyst to represent the actual spatial uncertainty as closely as possible, or quantified following an inverse approach using indirect measurement data (see, for example, Faes (2017) and Faes *et al.* (2017)). Also, Faes *et al.* (2019c) applied this method successfully in the modeling of spatial uncertainty in the thickness of parts produced via additive manufacturing. Similarly, van Mierlo *et al.* (2019) applied inverse distance weighting interpolation successfully to quantify stress–strain curve uncertainty in advanced constitutive material models. Finally, Faes and Moens (2020b) recently extended the method toward the modeling and simulation of interval fields taking cross-interdependence into account, in analogy to simulating cross-correlated random fields. A good example of the application of these concepts in realistic FE models is given in Faes *et al.* (2019c). The main drawback of this method is that it requires an *a priori* definition of the control point locations. While in some cases, such a definition can be intuitive (e.g. in the case of localized effects), this precludes the definition of a single measure for spatial dependence similar to a correlation length in random fields.

Local interval field decomposition (LIFD) was introduced in Imholz *et al.* (2015a, b) and also starts from the explicit formulation of the interval field, as introduced in equation [6.6]. The spatial dependence of the interval field realizations, evaluated at two separate locations in Ω, is limited by imposing an upper bound to the maximally occurring gradients of these realizations. Specifically, four global non-deterministic parameters are defined to bound all realizations $\boldsymbol{x}_j(\mathbf{r})$ of the interval field $\boldsymbol{x}^{\boldsymbol{I}}(\mathbf{r})$

$$\underline{\mu}_x \leq \mu_{x^I} \leq \overline{\mu}_x \quad [6.9a]$$

$$\forall \mathbf{r} \in \Omega : \boldsymbol{x}_j(\mathbf{r}) - \mu_{x^I} \leq \boldsymbol{s}_{x,max} \quad [6.9b]$$

$$\forall \mathbf{r} \in \Omega : \mu_{x^I} - \boldsymbol{x}_j(\mathbf{r}) \geq -\boldsymbol{s}_{x,max} \quad [6.9c]$$

$$-\frac{\partial x}{\partial \mathbf{r}}\Big|_{max} \leq \frac{\partial \boldsymbol{x}_j(\mathbf{r})}{\partial \mathbf{r}} \leq \frac{\partial x}{\partial \mathbf{r}}\Big|_{max} \quad [6.9d]$$

with $\boldsymbol{s}_{x,max}$ the maximum absolute value of the deviation from the mean value. In practice, this is achieved by defining the base functions $\boldsymbol{\psi}_i(\mathbf{r})$ as identically shaped, piecewise second-order polynomial functions for each separate element, which are located at the element midpoints (Imholz *et al.* 2015a, b). Recently also, a method to construct such base functions from a limited set of measurement data was introduced by Imholz *et al.* (2018). The LIFD method provides the analyst with an intuitive tool to model the spatial dependency of the interval field using a limited set of intuitive, globally defined parameters. However, a major disadvantage is computational cost corresponding to the propagation of the resulting interval field, as the dimension of the space spanned by the uncertain input parameters is equal to the number of elements in the model. This is particularly problematic in the case of industrially sized FE models containing up to millions of DOFs.

6.4.2. *Interval fields based on KL expansion*

An alternative pathway to include spatial or temporal dependence in an interval field model is by applying the KL expansion, as commonly applied in the simulation of random fields. Following the KL expansion method (see, for example, Sofi *et al.* (2015a, b), Sofi (2015) or Sofi and Muscolino (2015)), an interval field $\boldsymbol{x}^{\boldsymbol{I}}(\mathbf{r})$ is represented as:

$$\boldsymbol{x}^{\boldsymbol{I}}(\mathbf{r}) = \mu_{x^I}\left(1 + \sum_{i=1}^{N} \sqrt{\lambda_i}\boldsymbol{\psi}_i(\mathbf{r})\hat{e}_i^I\right) \quad [6.10]$$

with μ_{x^I} the mean of the interval field, and where λ_i and $\psi_i(r)$ are, respectively, the eigenvalues and eigenvectors of a deterministic, symmetric, non-negative, bounded function $\Gamma_B(\mathbf{r}_i, \mathbf{r}_j)$. The eigenvalues and eigenvectors correspond to the solution of the following eigenvalue problem:

$$\int_{\Omega} \Gamma_B(\mathbf{r}_i, \mathbf{r}_j)\psi_i(\mathbf{r}_i)d\mathbf{r} = \lambda_i \psi_i(\mathbf{r}) \qquad [6.11]$$

which is analogous to the homogeneous Fredholm equation of the second kind that is often encountered in the context of random fields (see, for example, Betz *et al.* (2014)). The main difference with random field analysis, however, lies in the definition of $\Gamma_B(\mathbf{r}_i, \mathbf{r}_j)$, which in this context is given as:

$$\Gamma_B(\mathbf{r}_i, \mathbf{r}_j) = mid\left(\boldsymbol{B}^{\boldsymbol{I}}(\mathbf{r}_i) \cdot \boldsymbol{B}^{\boldsymbol{I}}(\mathbf{r}_j)\right) = \frac{\boldsymbol{x}^{\boldsymbol{I}}(\mathbf{r}_i) \cdot \boldsymbol{x}^{\boldsymbol{I}}(\mathbf{r}_j)}{\mu^2_{x^I}} - 1 \qquad [6.12]$$

with $mid(\cdot)$ an operator that returns the midpoint of the interval between the brackets and $\boldsymbol{B}^{\boldsymbol{I}}(\mathbf{r})$ a dimensionless interval field with unit range (i.e. $\Delta \boldsymbol{B}(\boldsymbol{x}) < 1$). As such, when $mid(\cdot)$ is regarded as being analogous to a stochastic average operator for intervals, the function $\Gamma_B(\boldsymbol{r}_i, \boldsymbol{r}_j)$ may be considered as a non-probabilistic counterpart to the auto-correlation function of a random field. Indeed, it provides a measure of the dependence between the values of the dimensionless interval function $\boldsymbol{B}^I(\boldsymbol{r})$ at different locations $\boldsymbol{r}_i$ and $\boldsymbol{r}_j$. However, the optimal convergence rate of the KL expansion is lost by altering the formulation. Furthermore, only homogeneous dependence throughout the geometry can be defined in this way.

The $\hat{e}_i^I$ terms in equation [6.10] are extra unitary intervals, as defined in equation [6.5], and they are used to prevent the dependency phenomenon. Similarly to the explicit interval field decomposition method, the spatial dependence and uncertainty in the spatio-temporal uncertain property are represented separately by $\sqrt{\lambda_i}\psi_i(\mathbf{r})$ and the extra-unitary intervals $\hat{e}_i^I$, respectively. As such, commonly applied propagation methods, as discussed in section 6.2, can be applied. Obviously, this method integrates best with the improved interval analysis via the extra-unitary interval method, as presented by Sofi *et al.* (2019). Other recent developments (e.g. such as presented in Xia and Wang (2018)) in this direction include the application of sampling methods or Chebyshev polynomial expansions to propagate interval fields in time domain simulations. The method has been successfully applied to the Euler–Bernoulli beam analysis (Sofi 2015), Timoshenko beams (Sofi and Muscolino 2015) and uncertain structural free vibration analysis (Feng *et al.* 2019). A comparison of this method with more traditionally applied random field analysis has been provided by Sofi (2015). So far, published applications of the method seem to be limited to small-scale problems. Nonetheless, Sofi *et al.* (2019) recently integrated this method into commercial FE code to enable computing with realistic FE models, which should accelerate practical applicability.

6.4.3. *Interval fields based on convex descriptors*

An alternative approach to model interval fields (or processes) was introduced by Jiang *et al.* (2016) in the context of time domain uncertainty. At the core of this idea lies an auto-dependence function $\mathbf{\Gamma}$ to model spatial and/or temporal dependence of $x(\mathbf{r}_i)$ and $x(\mathbf{r}_j)$ according to an ellipsoidal model:

$$\mathbf{\Gamma}(\mathbf{r}_i, \mathbf{r}_j) = \cos(\theta)\sin(\theta)(l_j^2 - l_i^2) \quad [6.13]$$

with $\theta \in [-45\,\text{deg}; 45\,\text{deg}]$ an angle and l_1 and l_2 the half lengths of the major and minor axes of the ellipse. Note that in Jiang *et al.* (2016) and Ni and Jiang (2020), for example, the authors denote this function as an auto-correlation function. Via this approach, the admissible set $\mathcal{D}$ of two jointly dependent interval variables, located throughout the domain Ω, $x^I(\mathbf{r}_1)$ and $x^I(\mathbf{r}_2)$, can be analytically expressed as:

$$\mathcal{D} = \left\{ \begin{bmatrix} x(\mathbf{r}_1) \\ x(\mathbf{r}_2) \end{bmatrix} \middle| \begin{bmatrix} \hat{x}(\mathbf{r}_1) \\ \hat{x}(\mathbf{r}_2) \end{bmatrix}^T \begin{bmatrix} \mathbf{\Gamma}(\mathbf{r}_1, \mathbf{r}_1) & \mathbf{\Gamma}(\mathbf{r}_1, \mathbf{r}_2) \\ \mathbf{\Gamma}(\mathbf{r}_2, \mathbf{r}_1) & \mathbf{\Gamma}(\mathbf{r}_2, \mathbf{r}_2) \end{bmatrix} \begin{bmatrix} \hat{x}(\mathbf{r}_1) \\ \hat{x}(\mathbf{r}_2) \end{bmatrix} \right\} \quad [6.14]$$

with $\hat{x}(\mathbf{r}_i) = x(\mathbf{r}_i) - \mu_x(\mathbf{r}_i)$ as described in Ni and Jiang (2020). In the case where $\mathcal{D}$ is defined between all intervals that are distributed throughout Ω, their jointly admissible values are bounded. Hence, $\mathcal{D}$ is a good measure to bound the spatial dependence between $x^I(\mathbf{r}_1)$ and $x^I(\mathbf{r}_2)$. Another, unexplored pathway to model such admissible sets is to base it on the method presented by the authors in Faes and Moens (2019b).

As shown by Li *et al.* (2018), the formulation illustrated in equation [6.13] should be determined by available data. In this work, the authors propose to define a minimum volume ellipse that encloses the available data. This ellipse estimation is performed between all $x(\mathbf{r}_i)$ and $x(\mathbf{r}_j)$ that are available. Then, based on equation [6.13], a point-wise auto-dependence function can be estimated for the interval field. Based on definitions of the differentiability and integrability of this interval field concept, Li *et al.* (2019b) show how these concepts can be applied to compute the displacement, velocity and acceleration responses of a linear dynamical system, subjected to an interval process-valued base excitation. Hereto, they show that $\mathbf{\Gamma}(\mathbf{r}_i, \mathbf{r}_j)$ (equation [6.13]) can be directly translated to the radius of these responses of the system. Furthermore, Jiang *et al.* (2019) recently extended this method to continuum dynamic problems, as encountered, for example, in finite element analysis. Finally, in Ni and Jiang (2020), the method was generalized to static finite element problems with p-adaptive mesh refinement, greatly increasing the practical applicability of the methods.

6.5. Conclusion

This chapter gives an overview of recent developments in the field of interval field modeling. Three large classes of interval field methods exist, namely those based on (1) the explicit formulation, (2) affine arithmetic and (3) convex-set descriptors. Explicit formulation-based interval fields allow for defining both local and global uncertainties throughout the model geometry, based on very intuitive yet heuristic metrics to define dependence. The methods based on affine arithmetic integrate well with the well-known KL expansion method, allowing for the application of a globally defined homogeneous auto-dependence function. Finally, the methods based on convex-set descriptors are very flexible due to the pair-wise definition of the auto-dependence, but such definition can become very cumbersome for very large FE models.

6.6. Acknowledgments

The Research Foundation Flanders is gratefully acknowledged for the support of Matthias Faes under grant number 12P3519N and Maurice Imholz under project number G0C2218N.

6.7. References

Betz, W., Papaioannou, I., Straub, D. (2014). Numerical methods for the discretization of random fields by means of the Karhunen-Loeve expansion. *Comput. Methods Appl. Mech. Eng.*, 271, 109–129.

Bogaerts, L., Faes, M., Moens, D. (2019). A fast inverse approach for the quantification of set-theoretical uncertainty. *Proceedings of the 2019 IEEE Symposium Series on Computational Intelligence*. Xiamen, China.

Broggi, M., Faes, M., Patelli, E., Govers, Y., Moens, D., Beer, M. (2017). Comparison of Bayesian and interval uncertainty quantification: Application to the AIRMOD test structure. *Proceedings of the 2017 IEEE Symposium Series on Computational Intelligence*. Honolulu, USA.

Crespo, L.G., Giesy, D.P., Kenny, S.P. (2014). Interval predictor models with a formal characterization of uncertainty and reliability. *Proceedings of the IEEE Conference on Decision and Control*, February.

Dannert, M., Fau, A., Fleury, R., Broggi, M., Nackenhorst, U., Beer, M. (2018). A probability-box approach on uncertain correlation lengths by stochastic finite element method. *PAMM*, 18(1), e201800114.

De Munck, M., Moens, D., Desmet, W., Vandepitte, D. (2009). An efficient response surface based optimisation method for non-deterministic harmonic and transient dynamic analysis. *CMES*, 47(2), 119–166.

Degrauwe, D., Lombaert, G., Roeck, G.D. (2010). Improving interval analysis in finite element calculations by means of affine arithmetic. *Comput. and Struct.*, 88(3–4), 247–254.

Elishakoff, I. and Archaud, E. (2013). Modified Monte Carlo method for buckling analysis of nonlinear imperfect structures. *Arch. Appl. Mech.*, 83(9), 1327–1339.

Elishakoff, I. and Sarlin, N. (2016a). Uncertainty quantification based on pillars of experiment, theory, and computation. Part I: Data analysis. *Mech. Syst. Sig. Process.*, 74, 54–72.

Elishakoff, I. and Sarlin, N. (2016b). Uncertainty quantification based on pillars of experiment, theory, and computation. Part II: Theory and computation. *Mech. Syst. Sig. Process.*, 2, 74.

Faes, M. (2017). Interval methods for the identification and quantification of inhomogeneous uncertainty in Finite Element models. PhD Thesis, KU Leuven, Department of Mechanical Engineering, Belgium.

Faes, M. and Moens, D. (2017). Identification and quantification of spatial interval uncertainty in numerical models. *Comput. Struct.*, 192, 16–33.

Faes, M. and Moens, D. (2019a). Imprecise random field analysis with parametrized kernel functions. *Mech. Syst. Sig. Process.*, 134, 106334.

Faes, M. and Moens, D. (2019b). Multivariate dependent interval finite element analysis via convex hull pair constructions and the Extended Transformation Method. *Comput. Methods Appl. Mech. Eng.*, 347, 85–102.

Faes, M. and Moens, D. (2020a). Recent trends in the modeling and quantification of non-probabilistic uncertainty. *Arch. Comput. Methods Eng.*, 27(3), 633–671.

Faes, M. and Moens, D. (2020b). On auto and crossinterdependence in interval field finite element analysis. *Int. J. Numer. Methods Eng.*, 121(9).

Faes, M., Cerneels, J., Vandepitte, D., Moens, D. (2017). Identification and quantification of multivariate interval uncertainty in finite element models. *Comput. Methods Appl. Mech. Eng.*, 315, 896–920.

Faes, M., Broggi, M., Patelli, E., Govers, Y., Mottershead, J., Beer, M., Moens, D. (2019a). A multivariate interval approach for inverse uncertainty quantification with limited experimental data. *Mech. Syst. Sig. Process.*, 118, 534–548.

Faes, M., Sadeghi, J., Broggi, M., de Angelis, M., Patelli, E., Beer, M., Moens, D. (2019b). On the robust estimation of small failure probabilities for strong nonlinear models. *ASCE-ASME J. Risk and Uncert. Eng. Sys. Part B Mech. Eng.*, 5(4).

Faes, M., Sabyasachi, G.D., Moens, D. (2019c). Hybrid spatial uncertainty analysis for the estimation of imprecise failure probabilities in laser sintered PA-12 parts. *Comp. Math. Appl.*, 78(7), 2395–2406.

Feng, J., Li, Q., Sofi, A., Li, G., Wu, D., Gao, W. (2019). Uncertain structural free vibration analysis with non-probabilistic spatially varying parameters. *ASCE-ASME J. Risk Uncert. Eng. Syst., Part B Mech. Eng.*, 5(2).

Fujita, K. and Takewaki, I. (2011). An efficient methodology for robustness evaluation by advanced interval analysis using updated second-order Taylor series expansion. *Eng. Struct.*, 33(12), 3299–3310.

Hanss, M. (2005). *Applied Fuzzy Arithmetic*. Springer, Berlin, Heidelberg.

Imholz, M., Vandepitte, D., Moens, D. (2015a). Analysis of the effect of uncertain clamping stiffness on the dynamical behaviour of structures using interval field methods. *Appl. Mech. Mater.*, 807, 195–204.

Imholz, M., Vandepitte, D., Moens, D. (2015b). Derivation of an input interval field decomposition based on expert knowledge using locally defined basis functions. *Proceedings of the International Conference on Uncertainty Quantification in Computational Sciences and Engineering*, pp. 1–19.

Imholz, M., Vandepitte, D., Moens, D. (2018). Application of Interval Fields to fit experimental data on deepdrawn components. *Proceedings of the Joint ICVRAM ISUMA UNCERTAINTIES Conference*, University of Sao Paolo.

Jiang, C., Ni, B.Y., Liu, N.Y., Han, X., Liu, J. (2016). Interval process model and non-random vibration analysis. *J. Sound Vib.*, 373, 104–131.

Jiang, C., Li, J.W., Ni, B.Y., Fang, T. (2019). Some significant improvements for interval process model and non-random vibration analysis method. *Comput. Methods Appl. Mech. Eng.*, 357.

Khodaparast, H.H., Mottershead, J.E., Badcock, K.J. (2011). Interval model updating with irreducible uncertainty using the Kriging predictor. *Mech. Syst. Sig. Process.*, 25(4), 1204–1206.

Li, C., Chen, B., Peng, H., Zhang, S. (2017). Sparse regression Chebyshev polynomial interval method for nonlinear dynamic systems under uncertainty. *Appl. Math. Modell.*, 51, 505–525.

Li, J., Ni, B., Jiang, C., Fang, T. (2018). Dynamic response bound analysis for elastic beams under uncertain excitations. *J. Sound Vib.*, 422, 471–489.

Li, F., Liu, J., Wen, G., Rong, J. (2019a). Extending SORA method for reliability-based design optimization using probability and convex set mixed models. *Struct. Multidiscip. Optim.*, 59(4), 1163–1179.

Li, J., Jiang, C., Ni, B., Zhan, L. (2020). Uncertain vibration analysis based on the concept of differential and integral of interval process. *Int. J. Mech. Mat. Des*, 16(2), 225–244.

Liu, P.-L. and Der Kiureghian, A. (1991). Optimization algorithms for structural reliability. *Struct. Saf.*, 9(3), 161–177.

Liu, X.-X. and Elishakoff, I. (2019). Seismic risk analysis for reinforced concrete structures with both random and parallelepiped convex variables. *Struct. Infrastruct. Eng.*, 15(5), 618–633.

Manson, G. (2005). Calculating frequency response functions for uncertain systems using complex affine analysis. *J. Sound Vib.*, 288(3), 487–521.

Moens, D., De Munck, M., Desmet, W., Vandepitte, D. (2011). Numerical dynamic analysis of uncertain mechanical structures based on interval fields. *IUTAM Symposium on the Vibration Analysis of Structures with Uncertainties*. Springer, The Netherlands.

van Mierlo, C., Faes, M., Moens, D. (2019). Identification of visco-plastic material model parameters using interval fields. *Proceedings of the 3rd International Conference on Uncertainty Quantification in Computational Sciences and Engineering*, ECCOMAS.

Muhanna, R.L. and Mullen, R.L. (2001). Uncertanty in mechanics problems–interval–based approach. *J. Eng. Mech.*, 127(6), 557–566.

Muscolino, G. and Sofi, A. (2012). Stochastic analysis of structures with uncertain-but-bounded parameters via improved interval analysis. *Probab. Eng. Mech.*, 28, 152–163.

Muscolino, G. and Sofi, A. (2013). Bounds for the stationary stochastic response of truss structures with uncertain-but-bounded parameters. *Mech. Syst. Sig. Process.*, 37(1), 163–181.

Ni, B.Y. and Jiang, C. (2020). Interval field model and interval finite element analysis. *Comput. Methods Appl. Mech. Eng.*, 360.

Qiu, Z. and Elishakoff, I. (1998). Antioptimization of structures with large uncertain-but-non-random parameters via interval analysis. *Comput. Methods Appl. Mech. Eng.*, 152(3–4), 361–372.

Sadeghi, J., de Angelis, M., Patelli, E. (2020). Robust propagation of probability boxes by interval predictor models. *Struct. Saf.*, 82, 101889.

Sofi, A. (2015). Structural response variability under spatially dependent uncertainty: Stochastic versus interval model. *Probab. Eng. Mech.*, 42, 78–86.

Sofi, A. and Muscolino, G. (2015). Static analysis of Euler–Bernoulli beams with interval Young's modulus. *Comput. Struct.*, 156, 72–82.

Sofi, A. and Romeo, E. (2016). A novel interval finite element method based on the improved interval analysis. *Comput. Methods Appl. Mech. Eng.*, 311, 671–697.

Sofi, A. and Romeo, E. (2018). A unified response surface framework for the interval and stochastic finite element analysis of structures with uncertain parameters. *Probab. Eng. Mech.*, 54, 25–36.

Sofi, A., Muscolino, G., Elishakoff, I. (2015a). Natural frequencies of structures with interval parameters. *J. Sound Vib.*, 347, 79–95.

Sofi, A., Muscolino, G., Elishakoff, I. (2015b). Static response bounds of Timoshenko beams with spatially varying interval uncertainties. *Acta Mech.*, 226(11), 3737–48.

Sofi, A., Romeo, E., Barrera, O., Cocks, A. (2019). An interval finite element method for the analysis of structures with spatially varying uncertainties. *Adv. Eng. Software*, 128, 1–19.

Vanmarcke, E.H. and Grigoriu, M. (1983). Stochastic finite element analysis of simple beams. *J. Eng. Mech.*, 109(5), 1203–1214.

Verhaeghe, W., Desmet, W., Vandepitte, D., Moens, D. (2013a). Interval fields to represent uncertainty on the output side of a static FE analysis. *Comput. Methods Appl. Mech. Eng.*, 260, 50–62.

Verhaeghe, W., Elishakoff, I., Desmet, W., Vandepitte, D., Moens, D. (2013b). Uncertain initial imperfections via probabilistic and convex modeling: Axial impact buckling of a clamped beam. *Comp. Struct.*, 121, 1–9.

Wang, X., Elishakoff, I., Qiu, Z. (2008). Experimental data have to decide which of the nonprobabilistic uncertainty descriptions convex modeling or interval analysis to utilize. *J. Appl. Mech.*, 75(4), 41018.

Wang, L., Xiong, C., Wang, X., Xu, M., Li, Y. (2018). A dimension-wise method and its improvement for multidisciplinary interval uncertainty analysis. *Appl. Math. Modell.*, 59, 680–695.

Wu, D. and Gao, W. (2017). Uncertain static plane stress analysis with interval fields. *Int. J. Numer. Methods Eng.*, 110(13), 1272–1300.

Wu, J., Luo, Z., Zhang, Y., Zhang, N., Chen, L. (2013). Interval uncertain method for multibody mechanical systems using Chebyshev inclusion functions. *Int. J. Numer. Methods Eng.*, 95(7), 608–630.

Xia, B. and Wang, L. (2018). Non-probabilistic interval process analysis of time-varying uncertain structures. *Eng. Struct.*, 175, 101–112.

Zhan, J., Luo, Y., Zhang, X., Kang, Z. (2020). A general assessment index for non-probabilistic reliability of structures with bounded field and parametric uncertainties. *Comput. Methods Appl. Mech. Eng.*, 366.

Zhu, L., Elishakoff, I., Starnes Jr., J.H. (1996). Derivation of multi-dimensional ellipsoidal convex model for experimental data. *Math. Comput. Modell.*, 24(2), 103–114.

7

Convex Polytopic Models for the Static Response of Structures with Uncertain-but-bounded Parameters

7.1. Introduction

The accurate estimation of the static response of structures is crucial to the design and analysis of structures in many engineering applications. Conventionally, the numerical analysis of static response problems is usually performed for specified structural parameters and loading conditions. However, uncertainties originating from a lack of knowledge, manufacturing deviations, measurement errors and so on are inevitable in structures (Wang and Xiong 2019; Qiu and Jiang 2021). Generally speaking, the uncertainties, which may be present in the geometry, material properties, external loads or boundary conditions of structures (Qiu and Liu 2020; Wang and Liu 2020), continue to affect the design and operating performance of structures. For a proper performance assessment, these uncertainties must be taken into account appropriately (Fujita and Takewaki 2011).

It has been recognized that there are three main approaches to quantify the uncertainties depending on their nature and extent, i.e. probabilistic method, non-probabilistic method and fuzzy theory (Pantelides and Ganzerli 2001; Qiu and Lyu 2020). As the earliest method in the analysis of structures with uncertain parameters, the probabilistic method describes the uncertainties by the probability density distribution functions. Unfortunately, a wealth of data about the random variables is required to evaluate a probability, which is often unavailable accurately, especially when the number of samples is limited (Guo *et al.* 2008). In fuzzy theory, the uncertainties are described by the fuzzy membership functions, while the

---

Chapter written by Zhiping QIU and Nan JIANG.

membership function depends on the designers' experience (Ben-Haim and Elishakoff 1990). Thus, for cases where much of the information about uncertainties for either the probability density distribution function or the fuzzy membership function is unavailable, the non-probabilistic method, where only the bounds of the uncertainties are required, can be used most conveniently.

In view of the non-probabilistic method, the original papers were written by Ben-Haim and Elishakoff in the early 1990s (Ben-Haim and Elishakoff 1990; Elishakoff *et al.* 1994). Subsequently, it has demonstrated great superiority, arising from a lack of data in the last three decades, with Qiu (Qiu and Elishakoff 1998), Muhanna (Muhanna and Mullen 1999) and Ganzerli and Pantelides (2000; Pantelides and Ganzerli 2001) making great contributions. In fact, the non-probabilistic model is always used to deal with uncertain parameters in the convex domain, including three typical models, namely the interval model, the ellipsoid model and the convex polytopic model (Jalivand-Nejad *et al.* 2016), as shown in Figure 7.1.

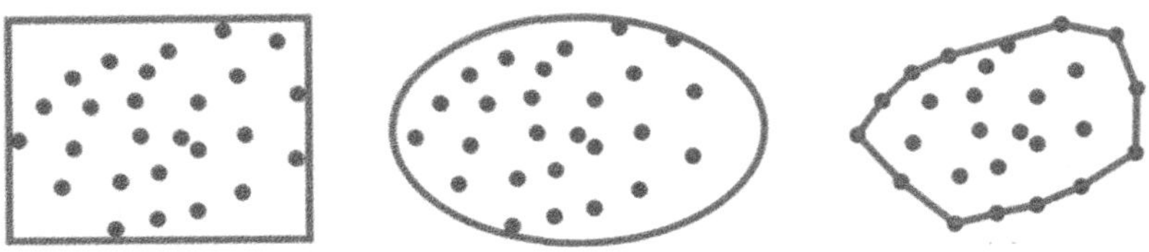

(a) Interval model (b) Ellipsoid model (c) Convex polytopic model

Figure 7.1. *Three typical two-dimensional convex models. For a color version of this figure, see www.iste.co.uk/challamel/mechanics3.zip*

For the static response problems in structures, the main aim is to compute the bounds of the static displacements when the uncertain parameters are constrained to belong to the convex domain. Among the three typical models, the interval model has attracted a great deal of attention due to its simplicity and convenience (Fujita and Takewaki 2011). The Oettli–Prager criterion was put forward initially, describing the full characterization of the solution set for linear interval equations (Oettli and Prager 1964). Rao (1984; Rao and Chen 1998) investigated a variety of alternative approximate methods, and various methods have been developed based on the perturbation theory and the Taylor's series expansion (Qiu and Elishakoff 1998; McWilliams 2001). Afterwards, the subinterval perturbation method (Zhou *et al.* 2006), the interval finite element method (Muhanna *et al.* 2007), the dimension-wise method (Xu and Qiu 2014) and the Bayesian collocation method (Liu *et al.* 2019) have been put forward successively. The vertex method was first proposed by Dong (Dong and Shah 1987), and then Qiu (Qiu *et al.* 2007; Qiu and Lv 2017) proved the vertex solution theorem to obtain the bounds of the static displacements. As for the optimization method, the problem of determining the bounds of the static response of structures with uncertain-but-bounded parameters is transformed into an optimization or programming problem, so that the problem can be solved by the algorithms in the optimization theory. Köylüoğlu (Köylüoğlu *et al.* 1995) first employed the linear programming to solve the discrete interval equations of structures. Guo (Guo *et al.* 2008) analyzed the extreme structural response of structures via linear mixed 0-1 programming. Li (Li *et al.* 2017) obtained the static response of structures by the difference of convex functions algorithm for quadratic programming problems. Luo (Luo *et al.* 2020) proposed an adjoint-based optimization method to predict the static response of structures.

In terms of the ellipsoid model, it is assumed that the uncertain parameter lies within a multidimensional ellipsoid (Kanno and Takewaki 2006). Qiu (Qiu 2003) performed a comparison of the static response of structures between the ellipsoid model and the interval model.

As a result, these studies may have two aspects of deficiencies, which should be taken into consideration. On one hand, existing uncertain parameters are usually dependent on each other in practice (Jiang *et al.* 2011; Sriramula and Chryssanthopoulos 2013; Jiang *et al.* 2014). The interval model neglects the correlation between the uncertain parameters considering them independent, and the ellipsoid model simplifies the correlation as a simple formula. However, it is worth noting that the correlation between the uncertain parameters is usually complex and difficult to be described as a unified form. Due to the simplicity of geometry, the interval model and the ellipsoid model inevitably bring about the problem of expansion in quantification. They may generate a larger range of the convex domain than the convex polytopic model, and some extra areas are often enveloped, which leads to the overestimation phenomenon. Compared with them, the convex polytopic model does not express the correlation as a simple formula with a unified form, but employs the minimum convex hull to represent uncertainties, which can achieve a tighter convex set and less redundant areas. On the other hand, although the smallest interval and ellipsoid containing the given experimental data can be derived to produce lower overestimation (Wang *et al.* 2008), they are usually not parallel to the global coordinate system. In the subsequent analysis, they first need to be transformed into those parallel to the global coordinate system, increasing the computational costs.

The convex polytopic model of the uncertain parameters is used to describe the uncertainty systems that first appeared in the field of robust control (Horisberger and Belanger 1976). The range of the uncertain parameters was only acquired by the vertices of the convex polyhedrons in parameter space, so as to evaluate the stability of the system. After 30 years of development, there have been a great number of research achievements on the convex polytopic model in robust control (de Oliveira *et al.* 2002; Mo 2011). Nevertheless, so far, there have been scarce studies on the convex polytopic model for the static response of structures. Inspired by the literature, whether we can obtain the bounds of the static displacements by the convex polytopic model is a challenging issue.

On the basis of these reasons, the main contribution of this chapter is to propose a novel convex polytopic model for the static response of structures with uncertain-but-bounded parameters, and its vertex solution theorem to predict the exact bounds of the static displacements. The proposed method can be applied as a powerful tool in static response analysis with uncertainties.

The remainder of this chapter is structured as follows. The convex polytopic model for the static response of structures with uncertain-but-bounded parameters is presented in section 7.2. In section 7.3, the convex polytopic model for the static response is transformed into two linear programming problems and the static displacement bound estimation is transformed into solving the optimal solutions of the linear programming problems. Then, the vertex solution theorem of the convex polytopic model for the static response is put forward in section 7.4, and the bounds of the static response can be obtained exactly. In section 7.5, the vertex solution theorem of the interval model is reviewed for the sake of comparison. Three numerical examples are accomplished to demonstrate the accuracy of the proposed method in section 7.6, and we finally conclude the chapter with a brief discussion.

7.2. Problem statements

The static response equation of structures in the finite element method can be expressed as

$$\mathbf{Ku} = \mathbf{f} \tag{7.1}$$

where $\mathbf{K} = (k_{ij})$ is the $n \times n$-dimensional structural stiffness matrix, $\mathbf{u} = (u_i)$ is the n-dimensional static displacement vector and $\mathbf{f} = (f_i)$ is the n-dimensional external load vector.

In practical engineering, it must be known that some inherent dispersions or uncertainties inevitably exist in both structural parameters and external loads. In some cases, the physical properties and geometric parameters of structures cannot be measured or manufactured exactly, and the load conditions cannot be obtained exactly either. As a result, the structural stiffness matrix $\mathbf{K}$ and the external load vector $\mathbf{f}$ are uncertain. Nevertheless, it should be realized that these uncertainties usually vary in a certain range, respectively, also called the uncertain-but-bounded parameters.

Under these circumstances, it is a common practice to seek the smallest closed and bounded convex set enclosing all possible values of the structural stiffness matrix $\mathbf{K}$ and the external load vector $\mathbf{f}$ to quantify the uncertain-but-bounded parameters, respectively, of which the frequently used models are the interval model and the ellipsoid model.

Based on the definition of a convex set, the uncertain-but-bounded structural stiffness matrix $\mathbf{K}$ and the uncertain-but-bounded external load vector $\mathbf{f}$ can be represented by the convex combinations in the following forms, respectively.

$$\mathbf{K} = \mathbf{K}(\boldsymbol{\alpha}) = \sum_{p=1}^{m_1} \alpha_p \mathbf{K}_p, \quad \sum_{p=1}^{m_1} \alpha_p = 1, \quad \alpha_p \geq 0 \qquad [7.2]$$

and

$$\mathbf{f} = \mathbf{f}(\boldsymbol{\beta}) = \sum_{q=1}^{m_2} \beta_q \mathbf{f}_q, \quad \sum_{q=1}^{m_2} \beta_q = 1, \quad \beta_q \geq 0 \qquad [7.3]$$

where $\mathbf{K}_p\ (p = 1,2,\cdots,m_1)$ denote the known structural stiffness vertex matrices which determine the range of the structural stiffness, $\mathbf{f}_q\ (q = 1,2,\cdots,m_2)$ denote the known external load vertex vectors, determining the range of the external load, and $\boldsymbol{\alpha} = (\alpha_p)$ and $\boldsymbol{\beta} = (\beta_q)$ denote the convex combination coefficient vectors, respectively.

As a result, the structural stiffness matrix $\mathbf{K} = \mathbf{K}(\boldsymbol{\alpha})$ and the external load vector $\mathbf{f} = \mathbf{f}(\boldsymbol{\beta})$ are uncertain but ranging inside the following convex polyhedrons, respectively.

$$\mathbf{K}^S(\boldsymbol{\alpha}) = \left\{ \mathbf{K} : \mathbf{K}(\boldsymbol{\alpha}) = \sum_{p=1}^{m_1} \alpha_p \mathbf{K}_p, \ \sum_{p=1}^{m_1} \alpha_p = 1, \ \alpha_p \geq 0 \right\} \qquad [7.4]$$

and

$$\mathbf{f}^S(\boldsymbol{\beta}) = \left\{ \mathbf{f} : \mathbf{f}(\boldsymbol{\beta}) = \sum_{q=1}^{m_2} \beta_q \mathbf{f}_q, \ \sum_{q=1}^{m_2} \beta_q = 1, \ \beta_q \geq 0 \right\} \qquad [7.5]$$

In such conditions, the structural stiffness $\mathbf{K}=\mathbf{K}(\boldsymbol{\alpha})$ and the external load $\mathbf{f}=\mathbf{f}(\boldsymbol{\beta})$ are called the convex polytopic uncertain parameters. Thus, the structural static response equation [7.1] with convex polytopic uncertain parameters can be rewritten as

$$\mathbf{K}(\boldsymbol{\alpha})\mathbf{u}=\mathbf{f}(\boldsymbol{\beta}) \qquad [7.6]$$

When the structural stiffness matrix $\mathbf{K}=\mathbf{K}(\boldsymbol{\alpha})$ and the external load vector $\mathbf{f}=\mathbf{f}(\boldsymbol{\beta})$ are ranging inside the aforementioned convex polyhedrons [7.4] and [7.5], respectively, the changing region of the structural static response, namely the solution to equation [7.6], is a set, and this set is given as

$$\mathbf{U}=\left\{\mathbf{u}:\mathbf{Ku}=\mathbf{f},\ \mathbf{K}\in\mathbf{K}^S(\boldsymbol{\alpha}),\ \mathbf{f}\in\mathbf{f}^S(\boldsymbol{\beta})\right\} \qquad [7.7]$$

However, generally speaking, it is extremely difficult to compute the solution set [7.7] exactly, since the solution set $\mathbf{U}$ may be of a complicated geometric shape, which may be a non-convex polyhedron (Moore *et al.* 2009). This is the difficulty of solving structural static response problems with uncertainties. Nevertheless, by taking this into account, we can determine the maximum and minimum values, or the upper and lower bounds of the structural static displacement set [7.7]. That is,

$$\mathbf{u}_{\min}=\underline{\mathbf{u}}\leq\mathbf{u}\leq\overline{\mathbf{u}}=\mathbf{u}_{\max} \qquad [7.8]$$

in which

$$\mathbf{u}_{\max}=\overline{\mathbf{u}}=\left(u_{i\max}\right)=\left(\overline{u_i}\right),\quad \mathbf{u}_{\min}=\underline{\mathbf{u}}=\left(u_{i\min}\right)=\left(\underline{u_i}\right),\quad i=1,2,\cdots,n \qquad [7.9]$$

where $\mathbf{u}_{\max}$ and $\mathbf{u}_{\min}$ are the maximum and minimum values of the structural static displacements, respectively.

The above proposed model is called the convex polytopic model for the static response of structures with uncertain-but-bounded parameters.

7.3. Analysis and solution of the convex polytopic model for the static response of structures

Considering the structural static response equation [7.6] with convex polytopic uncertain parameters, for any structural stiffness matrix $\mathbf{K}$ and external load vector $\mathbf{f}$ ranging inside the convex polyhedrons [7.4] and [7.5], respectively, the corresponding structural static displacement vector $\mathbf{u}$ can be represented by subtracting the non-negative vector $\mathbf{u}''$ from another non-negative vector $\mathbf{u}'$, i.e.

$$\mathbf{u} = \mathbf{u}' - \mathbf{u}'' \tag{7.10}$$

where $\mathbf{u}', \mathbf{u}'' \geq 0$. Here, $\mathbf{u}'$ does not mean the derivative or the transpose of $\mathbf{u}$. $\mathbf{u}''$ is also not the case.

Obviously, for any large enough $\mathbf{u}'' \geq 0$, $\mathbf{u}'$ which satisfies $\mathbf{u}' = \mathbf{u}'' + \mathbf{u}$ can ensure that $\mathbf{u}' \geq 0$. Therefore, countless $\mathbf{u}'$ and $\mathbf{u}''$ exist, satisfying equation [7.10] and $\mathbf{u}', \mathbf{u}'' \geq 0$.

Substituting equation [7.10] into equation [7.6] yields

$$\mathbf{K}(\mathbf{u}' - \mathbf{u}'') = \mathbf{f} \tag{7.11}$$

in the form of a matrix as

$$(\mathbf{K} \quad -\mathbf{K})\begin{pmatrix} \mathbf{u}' \\ \mathbf{u}'' \end{pmatrix} = \mathbf{f} \tag{7.12}$$

Let

$$\mathbf{A} = (\mathbf{K} \quad -\mathbf{K}), \quad \mathbf{x} = \begin{pmatrix} \mathbf{u}' \\ \mathbf{u}'' \end{pmatrix}, \quad \mathbf{b} = \mathbf{f} \tag{7.13}$$

According to equation [7.13], $\mathbf{A}$ is the $n \times 2n$-dimensional matrix, $\mathbf{x}$ is the $2n$-dimensional non-negative vector and $\mathbf{b}$ is the n-dimensional vector. Moreover, $\mathbf{A}$ and $\mathbf{b}$ belong to the following convex polyhedrons, respectively.

$$\mathbf{A}^S(\boldsymbol{\alpha}) = \left\{ \mathbf{A} : \mathbf{A}(\boldsymbol{\alpha}) = \sum_{p=1}^{m_1} \alpha_p \mathbf{A}_p, \ \sum_{p=1}^{m_1} \alpha_p = 1, \ \alpha_p \geq 0 \right\} \tag{7.14}$$

and

$$\mathbf{b}^S(\boldsymbol{\beta}) = \left\{ \mathbf{b} : \mathbf{b}(\boldsymbol{\beta}) = \sum_{q=1}^{m_2} \beta_q \mathbf{b}_q, \ \sum_{q=1}^{m_2} \beta_q = 1, \ \beta_q \geq 0 \right\} \tag{7.15}$$

where

$$\mathbf{A}_p = \begin{pmatrix} \mathbf{K}_p & -\mathbf{K}_p \end{pmatrix}, \quad \mathbf{b}_q = \mathbf{f}_q, \quad p = 1, 2, \cdots, m_1, \quad q = 1, 2, \cdots, m_2 \qquad [7.16]$$

Then, the structural static response equation [7.12] can be rewritten as

$$\mathbf{A}\mathbf{x} = \mathbf{b} \qquad [7.17]$$

In the linear equations [7.17], since the number of equations is less than the number of unknowns, there are numerous solutions $\mathbf{x}$ in total.

Next, we take the maximum and minimum values of the vector $\mathbf{x}$ into account. The maximum $\mathbf{x}_{\max}$ and minimum $\mathbf{x}_{\min}$ are given by

$$\mathbf{x}_{\max} = \overline{\mathbf{x}} = \left(x_{j\max}\right) = \left(\overline{x_j}\right), \quad \mathbf{x}_{\min} = \underline{\mathbf{x}} = \left(x_{j\min}\right) = \left(\underline{x_j}\right), \quad j = 1, 2, \cdots, 2n \qquad [7.18]$$

It is noteworthy that the vector $\mathbf{x}$ is only maximum when each component in $\mathbf{x}$ is maximum at the same time. Similarly, when each component in $\mathbf{x}$ is minimum at the same time, the vector $\mathbf{x}$ is minimum.

As a result, we can transform the problem of solving equation [7.18] into calculating the maximum $y_{\max}$ and minimum $y_{\min}$ of the scalar $y = \sum_{j=1}^{2n} x_j$. That is,

$$\begin{aligned} y_{\max} &= x_{1\max} + x_{2\max} + \cdots + x_{2n\max} = (1, 1, \cdots, 1)\left(x_{1\max}, x_{2\max}, \cdots, x_{2n\max}\right)^{\mathrm{T}}, \\ y_{\min} &= x_{1\min} + x_{2\min} + \cdots + x_{2n\min} = (1, 1, \cdots, 1)\left(x_{1\min}, x_{2\min}, \cdots, x_{2n\min}\right)^{\mathrm{T}} \end{aligned} \qquad [7.19]$$

Let

$$\mathbf{e} = (1, 1, \cdots, 1)^{\mathrm{T}} \qquad [7.20]$$

Substituting equation [7.20] into equation [7.19] gives

$$y_{\max} = \mathbf{e}^{\mathrm{T}}\mathbf{x}_{\max} = \max\left\{\mathbf{e}^{\mathrm{T}}\mathbf{x}\right\}, \quad y_{\min} = \mathbf{e}^{\mathrm{T}}\mathbf{x}_{\min} = \min\left\{\mathbf{e}^{\mathrm{T}}\mathbf{x}\right\} \qquad [7.21]$$

Thus, equation [7.21], subject to equation [7.17], can be described in the following standard form of the two linear programming problems as

$$\begin{aligned} & \max\left\{\mathbf{e}^{\mathrm{T}}\mathbf{x}\right\} \quad \text{and} \quad \min\left\{\mathbf{e}^{\mathrm{T}}\mathbf{x}\right\} \\ & \text{s.t. } \mathbf{A}\mathbf{x} = \mathbf{b} \\ & \qquad \mathbf{A} \in \mathbf{A}^{S}, \quad \mathbf{b} \in \mathbf{b}^{S} \\ & \qquad \mathbf{x} \geq 0 \end{aligned} \qquad [7.22]$$

After the above process, the convex polytopic model for the static response of structures is transformed into two linear programming problems under the constraints of the convex polyhedrons. Due to the arbitrariness of the matrix $\mathbf{K}$ and the vector $\mathbf{f}$, namely the arbitrariness of the matrix $\mathbf{A}$ and the vector $\mathbf{b}$, we can obtain the maximum and minimum values of $\mathbf{e}^{\mathrm{T}}\mathbf{x}$ from the above two linear programming problems. Consequently, the upper and lower bounds of the structural static displacements with the convex polytopic uncertain parameters can be obtained. The following key lies in solving the above linear programming problems.

7.4. Vertex solution theorem of the convex polytopic model for the static response of structures

In this section, the vertex solution theorem of the convex polytopic model for exact bounds of the structural static displacements is given. We can use the linear programming theory to predict the upper and lower bounds of the structural static displacements accurately.

To solve the aforementioned linear programming problems [7.22], we first introduce an important theorem in linear programming theory as follows.

THEOREM 7.1.– The optimal solution to a linear programming problem is reached on the vertex of the feasible region.

PROOF.– Consider the standard form of the linear programming problem as

$$\begin{aligned} & \min\left\{\mathbf{e}^{\mathrm{T}}\mathbf{x}\right\} \\ & \text{s.t. } \mathbf{A}\mathbf{x} = \mathbf{b} \\ & \qquad \mathbf{x} \geq 0 \end{aligned} \qquad [7.23]$$

Now, we prove it by contradiction. Assuming that $\mathbf{x}^{*}$ is the optimal solution to the linear programming problem [7.23], but not the vertex of the feasible region of the problem [7.23], there exists $\lambda \in [0,1]$ for the feasible solutions $\mathbf{x}_1, \mathbf{x}_2 \neq \mathbf{x}^*$ such that

$$\boldsymbol{x}^* = \lambda \boldsymbol{x}_1 + (1-\lambda)\boldsymbol{x}_2 \qquad [7.24]$$

Pre-multiplying both sides of equation [7.24] by $\mathbf{e}^{\mathrm{T}}$ yields

$$\mathbf{e}^{\mathrm{T}}\boldsymbol{x}^* = \lambda \mathbf{e}^{\mathrm{T}}\boldsymbol{x}_1 + (1-\lambda)\mathbf{e}^{\mathrm{T}}\boldsymbol{x}_2 \qquad [7.25]$$

Since $\mathbf{x}^*$ is the optimal solution to the problem [7.23], we have

$$\mathbf{e}^{\mathrm{T}}\boldsymbol{x}^* < \mathbf{e}^{\mathrm{T}}\boldsymbol{x}_1, \quad \mathbf{e}^{\mathrm{T}}\boldsymbol{x}^* < \mathbf{e}^{\mathrm{T}}\boldsymbol{x}_2 \qquad [7.26]$$

Multiplying the first formula of equation [7.26] by λ, and the second one by $1-\lambda$, gives

$$\lambda \mathbf{e}^{\mathrm{T}}\boldsymbol{x}^* < \lambda \mathbf{e}^{\mathrm{T}}\boldsymbol{x}_1, \quad (1-\lambda)\mathbf{e}^{\mathrm{T}}\boldsymbol{x}^* < (1-\lambda)\mathbf{e}^{\mathrm{T}}\boldsymbol{x}_2 \qquad [7.27]$$

Adding the two formulas of equation [7.27], we obtain

$$\mathbf{e}^{\mathrm{T}}\boldsymbol{x}^* < \lambda \mathbf{e}^{\mathrm{T}}\boldsymbol{x}_1 + (1-\lambda)\mathbf{e}^{\mathrm{T}}\boldsymbol{x}_2 \qquad [7.28]$$

Obviously, equation [7.28] is inconsistent with equation [7.25]. Therefore, the optimal solution $\mathbf{x}^*$ is the vertex of the feasible region of the problem [7.23].

For the maximum value problem

$$\begin{aligned} &\max\left\{\mathbf{e}^{\mathrm{T}}\mathbf{x}\right\} \\ &\text{s.t. } \mathbf{A}\mathbf{x} = \mathbf{b} \\ &\quad\;\; \mathbf{x} \geq 0 \end{aligned} \qquad [7.29]$$

it can be transformed to the form of the minimum value problem as

$$\begin{aligned} &\min\left\{-\mathbf{e}^{\mathrm{T}}\mathbf{x}\right\} \\ &\text{s.t. } \mathbf{A}\mathbf{x} = \mathbf{b} \\ &\quad\;\; \mathbf{x} \geq 0 \end{aligned} \qquad [7.30]$$

Assuming that $\mathbf{x}^{**}$ is the optimal solution to the linear programming problem [7.30], but not the vertex of the feasible region of the problem [7.30], we can prove Theorem 7.1 by a process similar to equations [7.24]–[7.28].

According to Theorem 7.1, in the simplex $\mathbf{Ax}=\mathbf{b},\mathbf{x}\geq 0$, the optimal solutions $\max\{\mathbf{e}^{\mathrm{T}}\mathbf{x}\}$ and $\min\{\mathbf{e}^{\mathrm{T}}\mathbf{x}\}$ to the linear programming problem [7.22] are reached on the vertices of the set of the feasible solutions $\mathbf{x}$, i.e. the feasible region $\mathbf{D}=\{\mathbf{x}:\mathbf{Ax}=\mathbf{b},\mathbf{x}\geq 0\}$, respectively.

Considering the convex polyhedrons [7.14] and [7.15], which the matrix $\mathbf{A}$ and the vector $\mathbf{b}$ belong to, respectively, we select the vertex matrix $\mathbf{A}_p\,(p=1,2,\cdots,m_1)$ of the convex polyhedron [7.14] and the vertex vector $\mathbf{b}_q\,(q=1,2,\cdots,m_2)$ of the convex polyhedron [7.15] arbitrarily. Therefore, the following combined linear programming problems formed by combining the vertex matrices $\mathbf{A}_p\,(p=1,2,\cdots,m_1)$ and the vertex vectors $\mathbf{b}_q\,(q=1,2,\cdots,m_2)$ are established.

$$\begin{aligned}&\max\{\mathbf{e}^{\mathrm{T}}\mathbf{x}\}\quad\text{and}\quad\min\{\mathbf{e}^{\mathrm{T}}\mathbf{x}\}\\&\text{s.t. }\mathbf{A}_p\mathbf{x}=\mathbf{b}_q\\&\qquad\mathbf{x}\geq 0\\&\qquad p=1,2,\cdots,m_1\\&\qquad q=1,2,\cdots,m_2\end{aligned}\tag{7.31}$$

For any subproblem of the above combined linear programming problems [7.31], the maximum and minimum values of $\mathbf{e}^{\mathrm{T}}\mathbf{x}$ are reached on the vertices of the feasible region $\mathbf{D}=\{\mathbf{x}:\mathbf{A}_p\mathbf{x}=\mathbf{b}_q,\ \mathbf{x}\geq 0\}\,(p=1,2,\cdots,m_1,\ q=1,2,\cdots,m_2)$. Here, we mark the maximum and minimum values of the vertices of the feasible regions of the linear programming subproblems as

$$\begin{aligned}&\overline{\mathbf{x}}_{pq}=\max\{\mathbf{x}_{pq}:\mathbf{A}_p\mathbf{x}_{pq}=\mathbf{b}_q,\ \mathbf{x}_{pq}\geq 0\},\\&\underline{\mathbf{x}}_{pq}=\min\{\mathbf{x}_{pq}:\mathbf{A}_p\mathbf{x}_{pq}=\mathbf{b}_q,\ \mathbf{x}_{pq}\geq 0\},\\&p=1,2,\cdots,m_1,\quad q=1,2,\cdots,m_2\end{aligned}\tag{7.32}$$

Obviously, the simplexes

$$\mathbf{A}_p\mathbf{x}=\mathbf{b}_q,\quad \mathbf{x}\geq 0,\quad p=1,2,\cdots,m_1,\quad q=1,2,\cdots,m_2\tag{7.33}$$

can be viewed as a combination of the simplexes of a series of linear programming subproblems. It can be inferred that the maximum and minimum values of $\mathbf{e}^{\mathrm{T}}\mathbf{x}$ of the original linear programming problems [7.22] are reached on the vertices of the region generated by combining the feasible regions of all the linear programming subproblems.

As a result, the maximum of all the maximum values $\overline{\mathbf{x}}_{pq}$ of the linear programming subproblems can be represented as the maximum values of the original linear programming problems [7.22] and the minimum of all the minimum values $\underline{\mathbf{x}}_{pq}$ can be perceived as the minimum values of the original linear programming problems [7.22]. That is,

$$\begin{aligned}\mathbf{x}_{\max} &= \max\left\{\overline{\mathbf{x}}_{pq} : \mathbf{A}_p\overline{\mathbf{x}}_{pq} = \mathbf{b}_q,\ \overline{\mathbf{x}}_{pq} \geq 0,\ p=1,2,\cdots,m_1,\ q=1,2,\cdots,m_2\right\},\\ \mathbf{x}_{\min} &= \min\left\{\underline{\mathbf{x}}_{pq} : \mathbf{A}_p\underline{\mathbf{x}}_{pq} = \mathbf{b}_q,\ \underline{\mathbf{x}}_{pq} \geq 0,\ p=1,2,\cdots,m_1,\ q=1,2,\cdots,m_2\right\}\end{aligned} \qquad [7.34]$$

On this account, by considering the variable substitution formula [7.13], we have

$$\mathbf{x}_{\max} = \begin{pmatrix}\mathbf{x}'_{\max}\\ \mathbf{x}''_{\max}\end{pmatrix},\quad \mathbf{x}_{\min} = \begin{pmatrix}\mathbf{x}'_{\min}\\ \mathbf{x}''_{\min}\end{pmatrix} \qquad [7.35]$$

Thus, we can obtain the maximum and minimum values, or the upper and lower bounds of the structural static displacements

$$\mathbf{u}_{\max} = \overline{\mathbf{u}} = \mathbf{x}'_{\max} - \mathbf{x}''_{\min},\quad \mathbf{u}_{\min} = \underline{\mathbf{u}} = \mathbf{x}'_{\min} - \mathbf{x}''_{\max} \qquad [7.36]$$

7.5. Review of the vertex solution theorem of the interval model for the static response of structures

For the sake of comparison, the vertex solution theorem of the interval model for the static response of structures (Qiu and Lv 2017) is reviewed for estimating the structural static displacements.

In this method, the uncertain-but-bounded parameters in the static system of structures are described in the interval model to the interval parameters, where the parameters are considered to be independent of each other. Therefore, equation [7.1] becomes a linear equation, subject to the following constraint conditions:

$$\underline{\mathbf{K}} \leq \mathbf{K} \leq \overline{\mathbf{K}},\quad \underline{\mathbf{f}} \leq \mathbf{f} \leq \overline{\mathbf{f}} \qquad [7.37]$$

where the upper and lower bound matrices of the stiffness matrix $\mathbf{K}$ are, respectively,

$$\overline{\mathbf{K}}=\begin{bmatrix}\overline{k}_{11} & \overline{k}_{12} & \cdots & \overline{k}_{1n}\\ \overline{k}_{21} & \overline{k}_{22} & \cdots & \overline{k}_{2n}\\ \vdots & \vdots & \ddots & \vdots\\ \overline{k}_{n1} & \overline{k}_{n2} & \cdots & \overline{k}_{nn}\end{bmatrix},\quad \underline{\mathbf{K}}=\begin{bmatrix}\underline{k}_{11} & \underline{k}_{12} & \cdots & \underline{k}_{1n}\\ \underline{k}_{21} & \underline{k}_{22} & \cdots & \underline{k}_{2n}\\ \vdots & \vdots & \ddots & \vdots\\ \underline{k}_{n1} & \underline{k}_{n2} & \cdots & \underline{k}_{nn}\end{bmatrix} \qquad [7.38]$$

and the upper and lower bound vectors of the external load vector $\mathbf{f}$ are, respectively,

$$\overline{\mathbf{f}}=\left[\overline{f}_1,\overline{f}_2,\cdots,\overline{f}_n\right]^{\mathrm{T}},\quad \underline{\mathbf{f}}=\left[\underline{f}_1,\underline{f}_2,\cdots,\underline{f}_n\right]^{\mathrm{T}} \qquad [7.39]$$

Making use of the interval matrix notation in interval analysis, the inequality conditions [7.37] can be written as

$$\mathbf{K}\in\mathbf{K}^I=\left[\underline{\mathbf{K}},\overline{\mathbf{K}}\right]=\left(k_{ij}^I\right),\quad \mathbf{f}\in\mathbf{f}^I=\left[\underline{\mathbf{f}},\overline{\mathbf{f}}\right]=\left(f_i^I\right) \qquad [7.40]$$

Thus, equation [7.1], subject to the constraint conditions [7.40], can be expressed by the linear interval equation

$$\mathbf{K}^I\mathbf{u}=\mathbf{f}^I \qquad [7.41]$$

where $\mathbf{K}^I=\left(k_{ij}^I\right)$ is the $n\times n$ -dimensional interval stiffness matrix and $\mathbf{f}^I=\left(f_i^I\right)$ is the n -dimensional interval external load vector.

Based on the Cramer's rule of linear equations, the solutions to equation [7.41] are

$$u_i=\frac{D_i\left(k_{1j_1}k_{2j_2}\cdots f_{ij_i}\cdots k_{nj_n}\right)}{D\left(k_{1j_1}k_{2j_2}\cdots k_{ij_i}\cdots k_{nj_n}\right)},\quad i=1,2,\cdots,n \qquad [7.42]$$

and

$$\underline{k}_{ij}\le k_{ij}\le\overline{k}_{ij},\quad \underline{f}_i\le f_i\le\overline{f}_i,\quad i,j=1,2,\cdots,n \qquad [7.43]$$

where

$$D_i\left(k_{1j_1}k_{2j_2}\cdots f_{ij_i}\cdots k_{nj_n}\right)=\begin{vmatrix} k_{11} & \cdots & k_{1,i-1} & f_1 & k_{1,i+1} & \cdots & k_{1n} \\ & \vdots & & \vdots & & \vdots & \\ k_{n1} & \cdots & k_{n,i-1} & f_n & k_{n,i+1} & \cdots & k_{nn} \end{vmatrix},$$

$$D\left(k_{1j_1}k_{2j_2}\cdots k_{ij_i}\cdots k_{nj_n}\right)=\begin{vmatrix} k_{11} & k_{12} & \cdots & k_{1n} \\ k_{21} & k_{22} & \cdots & k_{2n} \\ \vdots & \vdots & \ddots & \vdots \\ k_{n1} & k_{n2} & \cdots & k_{nn} \end{vmatrix}\neq 0 \qquad [7.44]$$

Let $k_{1j_1}^v, k_{2j_2}^v, \cdots, k_{nj_n}^v$ and f_i^v be vertices of $k_{1j_1}, k_{2j_2}, \cdots, k_{nj_n}$ and f_i, respectively. We have the set of the extreme points

$$\Omega=\left\{k_{1j_1}^v, k_{2j_2}^v, \cdots, k_{nj_n}^v, f_i^v,\ j_1, j_2, \cdots, j_n, i=1,2,\cdots,n,\ v=1,2,\cdots,m\right\} \qquad [7.45]$$

The elements k_{ij} of the interval stiffness matrix $\mathbf{K}^I$ and the components f_i of the interval external load vector $\mathbf{f}^I$ can be represented as

$$k_{lj_l}=\sum_{v=1}^{m}\lambda_{lj_l}^v k_{lj_l}^v,\quad f_i=\sum_{v=1}^{m}\lambda_i^v f_i^v,\quad l=1,2,\cdots,n \qquad [7.46]$$

where

$$\sum_{v=1}^{m}\lambda_{lj_l}^v=1,\quad \sum_{v=1}^{m}\lambda_i^v=1,\quad \lambda_{lj_l}^v\geq 0,\quad \lambda_i^v\geq 0,\quad l=1,2,\cdots,n \qquad [7.47]$$

According to equation [7.42], the solution of equation [7.41], subject to the extreme points set Ω, can be given by

$$u_i^v=\frac{D_i\left(k_{1j_1}^v k_{2j_2}^v\cdots f_{ij_i}^v\cdots k_{nj_n}^v\right)}{D\left(k_{1j_1}^v k_{2j_2}^v\cdots k_{ij_i}^v\cdots k_{nj_n}^v\right)},\quad i=1,2,\cdots,n \qquad [7.48]$$

Suppose that the maximum values of u_i^v will be reached on the extreme points $k_{1j_1}^p, k_{2j_2}^p, \cdots, k_{nj_n}^p$ and $f_i^p\ (1\leq p\leq m)$. We have

$$u_i^p=u_{i\max}^v=\max\left\{u_i^v\right\} \qquad [7.49]$$

where

$$u_i^p = \frac{D_i\left(k_{1j_1}^p k_{2j_2}^p \cdots f_{ij_i}^p \cdots k_{nj_n}^p\right)}{D\left(k_{1j_1}^p k_{2j_2}^p \cdots k_{ij_i}^p \cdots k_{nj_n}^p\right)} \tag{7.50}$$

Equation [7.49], in other words, can be expressed in the form of inequality

$$u_i^p \geq u_i^v \tag{7.51}$$

Substituting equations [7.48] and [7.50] into inequality [7.51] gives

$$\frac{D_i\left(k_{1j_1}^p k_{2j_2}^p \cdots f_{ij_i}^p \cdots k_{nj_n}^p\right)}{D\left(k_{1j_1}^p k_{2j_2}^p \cdots k_{ij_i}^p \cdots k_{nj_n}^p\right)} \geq \frac{D_i\left(k_{1j_1}^v k_{2j_2}^v \cdots f_{ij_i}^v \cdots k_{nj_n}^v\right)}{D\left(k_{1j_1}^v k_{2j_2}^v \cdots k_{ij_i}^v \cdots k_{nj_n}^v\right)} \tag{7.52}$$

Both sides of the inequality [7.51] being multiplied by the quantity $D\left(k_{1j_1}^p k_{2j_2}^p \cdots k_{ij_i}^p \cdots k_{nj_n}^p\right) D\left(k_{1j_1}^v k_{2j_2}^v \cdots k_{ij_i}^v \cdots k_{nj_n}^v\right)$, gives

$$\begin{aligned} &D_i\left(k_{1j_1}^p k_{2j_2}^p \cdots f_{ij_i}^p \cdots k_{nj_n}^p\right) D\left(k_{1j_1}^v k_{2j_2}^v \cdots k_{ij_i}^v \cdots k_{nj_n}^v\right) \\ &\geq D_i\left(k_{1j_1}^v k_{2j_2}^v \cdots f_{ij_i}^v \cdots k_{nj_n}^v\right) D\left(k_{1j_1}^p k_{2j_2}^p \cdots k_{ij_i}^p \cdots k_{nj_n}^p\right) \end{aligned} \tag{7.53}$$

It is noteworthy that for any $k_{ij} \in \left[\underline{k}_{ij}, \overline{k}_{ij}\right] (i, j = 1, 2, \cdots, n)$, the quantity $D_i\left(k_{1j_1}^p k_{2j_2}^p \cdots f_{ij_i}^p \cdots k_{nj_n}^p\right) D\left(k_{1j_1}^v k_{2j_2}^v \cdots k_{ij_i}^v \cdots k_{nj_n}^v\right)$ is positive.

Both sides of the inequality [7.53], being multiplied by $\lambda_{1j_1}^v, \lambda_{2j_2}^v, \cdots, \lambda_{nj_n}^v$ and λ_i^v, gives

$$\begin{aligned} &\lambda_{1j_1}^v \lambda_{2j_2}^v \cdots \lambda_{nj_n}^v \lambda_i^v D_i\left(k_{1j_1}^p k_{2j_2}^p \cdots f_{ij_i}^p \cdots k_{nj_n}^p\right) D\left(k_{1j_1}^v k_{2j_2}^v \cdots k_{ij_i}^v \cdots k_{nj_n}^v\right) \\ &\geq \lambda_{1j_1}^v \lambda_{2j_2}^v \cdots \lambda_{nj_n}^v \lambda_i^v D_i\left(k_{1j_1}^v k_{2j_2}^v \cdots f_{ij_i}^v \cdots k_{nj_n}^v\right) D\left(k_{1j_1}^p k_{2j_2}^p \cdots k_{ij_i}^p \cdots k_{nj_n}^p\right) \end{aligned} \tag{7.54}$$

By taking summation with respect to v, we can obtain the following inequality:

$$\begin{aligned}&\sum_{v=1}^{m}\lambda_i^v D_i\left(k_{1j_1}^p k_{2j_2}^p \cdots f_{ij_i}^p \cdots k_{nj_n}^p\right)\\&\cdot D\left(\sum_{v=1}^{m}\lambda_{1j_1}^v k_{1j_1}^v \sum_{v=1}^{m}\lambda_{2j_2}^v k_{2j_2}^v \cdots \sum_{v=1}^{m}\lambda_{ij_i}^v k_{ij_i}^v \cdots \sum_{v=1}^{m}\lambda_{nj_n}^v k_{nj_n}^v\right)\\&\geq \sum_{v=1}^{m}\lambda_{ip_i}^v D\left(k_{1j_1}^p k_{2j_2}^p \cdots k_{ij_i}^p \cdots k_{nj_n}^p\right)\\&\cdot D_i\left(\sum_{v=1}^{m}\lambda_{1j_1}^v k_{1j_1}^v \sum_{v=1}^{m}\lambda_{2j_2}^v k_{2j_2}^v \cdots \sum_{v=1}^{m}\lambda_i^v f_{ij_i}^v \cdots \sum_{v=1}^{m}\lambda_{nj_n}^v k_{nj_n}^v\right)\end{aligned} \quad [7.55]$$

Substituting equations [7.46] and [7.47] into inequality [7.55] gives

$$\begin{aligned}&D_i\left(k_{1j_1}^p k_{2j_2}^p \cdots f_{ij_i}^p \cdots k_{nj_n}^p\right) D\left(k_{1j_1} k_{2j_2} \cdots k_{ij_i} \cdots k_{nj_n}\right)\\&\geq D\left(k_{1j_1}^p k_{2j_2}^p \cdots k_{ij_i}^p \cdots k_{nj_n}^p\right) D_i\left(k_{1j_1} k_{2j_2} \cdots f_{ij_i} \cdots k_{nj_n}\right)\end{aligned} \quad [7.56]$$

Both sides of the inequality [7.56], being divided by the quantity $D\left(k_{1j_1}^p k_{2j_2}^p \cdots k_{ij_i}^p \cdots k_{nj_n}^p\right) D\left(k_{1j_1} k_{2j_2} \cdots k_{ij_i} \cdots k_{nj_n}\right)$, gives

$$u_i^p = \frac{D_i\left(k_{1j_1}^p k_{2j_2}^p \cdots f_{ij_i}^p \cdots k_{nj_n}^p\right)}{D\left(k_{1j_1}^p k_{2j_2}^p \cdots k_{ij_i}^p \cdots k_{nj_n}^p\right)} \geq \frac{D_i\left(k_{1j_1} k_{2j_2} \cdots f_{ij_i} \cdots k_{nj_n}\right)}{D\left(k_{1j_1} k_{2j_2} \cdots k_{ij_i} \cdots k_{nj_n}\right)} = u_i \quad [7.57]$$

namely

$$u_{i\max}^v = u_i^p \geq u_i \quad [7.58]$$

In a similar manner, the following inequality can be deduced as

$$u_i^q = \frac{D_i\left(k_{1j_1}^q k_{2j_2}^q \cdots f_{ij_i}^q \cdots k_{nj_n}^q\right)}{D\left(k_{1j_1}^q k_{2j_2}^q \cdots k_{ij_i}^q \cdots k_{nj_n}^q\right)} \leq \frac{D_i\left(k_{1j_1} k_{2j_2} \cdots f_{ij_i} \cdots k_{nj_n}\right)}{D\left(k_{1j_1} k_{2j_2} \cdots k_{ij_i} \cdots k_{nj_n}\right)} = u_i \quad [7.59]$$

namely

$$u_{i\min}^v = u_i^q \geq u_i \quad [7.60]$$

where u_i^q is the minimum of u_i^v.

It can be seen from inequalities [7.57] and [7.59] that the maximum and minimum values, or the upper and lower bounds of the static displacement u_i of equation [7.41] will be reached on the vertex stiffness matrix $\mathbf{K}^v = \left(k_{ij}^v\right)$ and vertex external load vector $\mathbf{f}^v = \left(f_i^v\right)$.

7.6. Numerical examples

In this section, three numerical examples are provided to illustrate the effectiveness of the vertex solution theorem based on the convex polytopic model, including a two-step bar, a ten-bar truss and a plane frame. For comparison, the numerical results of the vertex solution theorem based on the convex polytopic model (CPM) and the interval model (IM) are given, respectively.

7.6.1. *Two-step bar*

First, consider the two-step bar shown in Figure 7.2. Two external loads f_1 and f_2 are applied at nodes 2 and 3, respectively. The stiffness terms of the two steps are k_1 and k_2, respectively. Here, the static displacements of nodes 2 and 3 are calculated, denoted by u_1 and u_2, respectively.

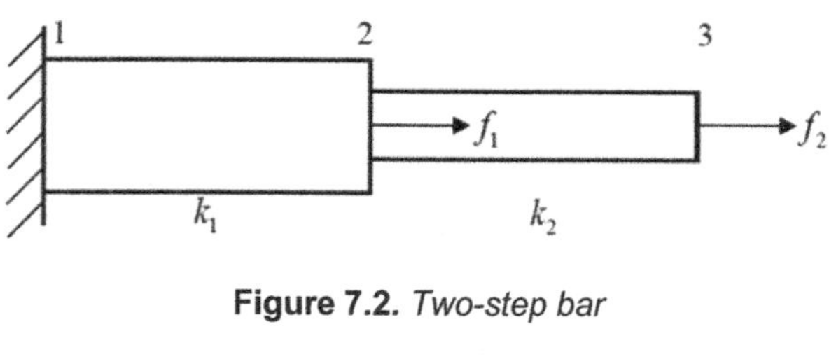

Figure 7.2. *Two-step bar*

The global static response equation in the finite element method is

$$\mathbf{Ku} = \mathbf{f} \tag{7.61}$$

where $\mathbf{K} = \begin{pmatrix} k_1 + k_2 & -k_2 \\ -k_2 & k_2 \end{pmatrix}$ is the global stiffness matrix, $\mathbf{u} = \begin{pmatrix} u_1 & u_2 \end{pmatrix}^{\mathrm{T}}$ is the static displacement vector and $\mathbf{f} = \begin{pmatrix} f_1 & f_2 \end{pmatrix}^{\mathrm{T}}$ is the global external load vector.

Due to unavoidable manufacturing errors and load deviations, the global stiffness matrix $\mathbf{K}$ and the global external load vector $\mathbf{f}$ are uncertain ranging inside the following convex polyhedrons, respectively

$$\mathbf{K} \in \mathbf{K}^S(\boldsymbol{\alpha}) = \left\{ \mathbf{K} : \mathbf{K} = \sum_{p=1}^{6} \alpha_p \mathbf{K}_p,\ \sum_{p=1}^{6} \alpha_p = 1,\ \alpha_p \geq 0,\ p = 1, 2, \cdots, 6 \right\} \quad [7.62]$$

and

$$\mathbf{f} \in \mathbf{f}^S(\boldsymbol{\beta}) = \left\{ \mathbf{f} : \mathbf{f} = \sum_{q=1}^{4} \beta_q \mathbf{f}_q,\ \sum_{q=1}^{4} \beta_q = 1,\ \beta_q \geq 0,\ q = 1, 2, 3, 4 \right\} \quad [7.63]$$

where $\mathbf{K}^S$ and $\mathbf{f}^S$ denote the set combined by the known stiffness vertex matrices $\mathbf{K}_p\ (p = 1, 2, \cdots, 6)$ and the set combined by the known external load vertex vectors $\mathbf{f}_q\ (q = 1, 2, 3, 4)$, respectively. The vertex matrices $\mathbf{K}_p\ (p = 1, 2, \cdots, 6)$ and the vertex vectors $\mathbf{f}_q\ (q = 1, 2, 3, 4)$ are, respectively

$$\begin{aligned} &\mathbf{K}_1 = \begin{pmatrix} 29.7 & -19.8 \\ -19.8 & 19.8 \end{pmatrix},\ \mathbf{K}_2 = \begin{pmatrix} 29.8 & -19.8 \\ -19.8 & 19.8 \end{pmatrix},\ \mathbf{K}_3 = \begin{pmatrix} 30.1 & -20 \\ -20 & 20 \end{pmatrix}, \\ &\mathbf{K}_4 = \begin{pmatrix} 30.3 & -20.2 \\ -20.2 & 20.2 \end{pmatrix},\ \mathbf{K}_5 = \begin{pmatrix} 30.2 & -20.2 \\ -20.2 & 20.2 \end{pmatrix},\ \mathbf{K}_6 = \begin{pmatrix} 29.9 & -20 \\ -20 & 20 \end{pmatrix} \end{aligned} \quad [7.64]$$

and

$$\mathbf{f}_1 = \begin{pmatrix} 49.5 \\ 29.7 \end{pmatrix},\ \mathbf{f}_2 = \begin{pmatrix} 50.5 \\ 29.7 \end{pmatrix},\ \mathbf{f}_3 = \begin{pmatrix} 50.5 \\ 30.3 \end{pmatrix},\ \mathbf{f}_4 = \begin{pmatrix} 49.5 \\ 30.3 \end{pmatrix} \quad [7.65]$$

The convex polytopic models of the stiffness terms k_1, k_2 and the external loads f_1, f_2 are shown in Figure 7.3(a) and (b), respectively.

The numerical results of the static displacements u_1 and u_2, obtained by the vertex solution theorem based on CPM and IM, are plotted in Figure 7.4, where the shadow region represents the solution domain determined by CPM and the hexagon surrounded by the blue lines denotes the bounds of the solution determined by IM.

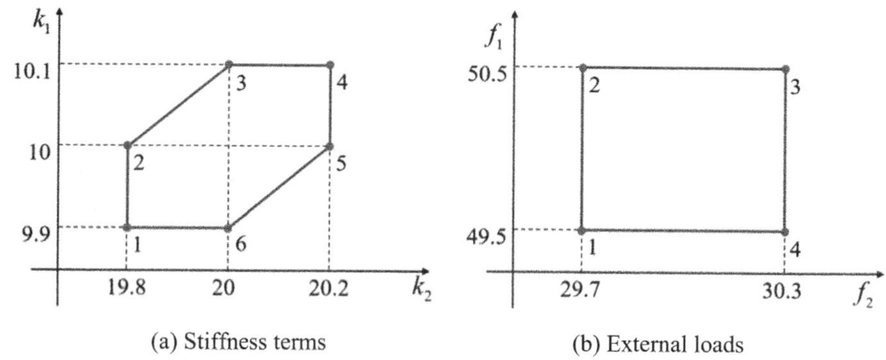

Figure 7.3. *Convex polytopic models of the stiffness terms and the external loads of the two-step bar. For a color version of this figure, see www.iste.co.uk/challamel/mechanics3.zip*

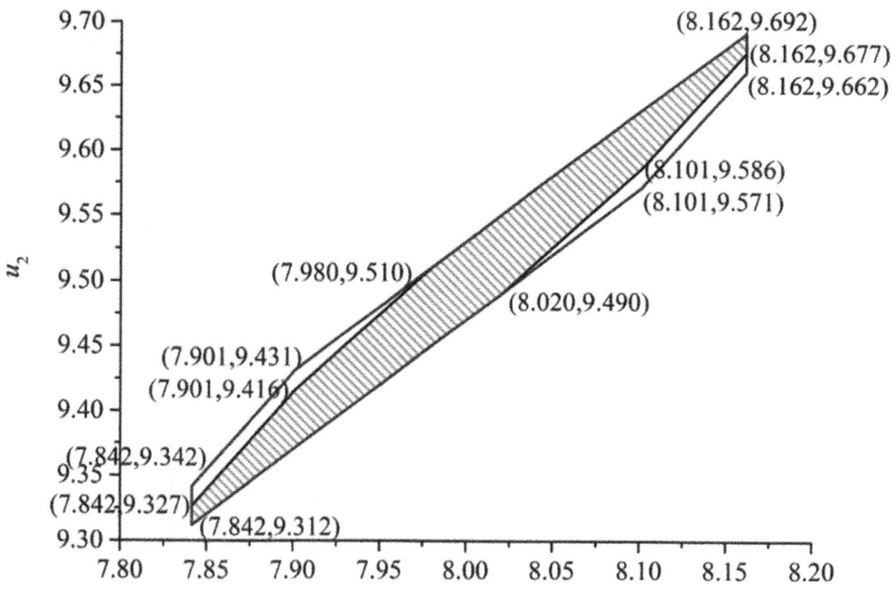

Figure 7.4. *Static displacements u_1 and u_2 of the vertex solution theorem, based on CPM and IM of the two-step bar. For a color version of this figure, see www.iste.co.uk/challamel/mechanics3.zip*

From Figure 7.4, it is shown that the shadow region is completely inside the hexagon surrounded by the blue lines. In other words, the bounds of the solution determined by IM enclose the solution domain determined by CPM completely. Furthermore, it is noteworthy that most of vertices of the solution domain determined by CPM are located on the bounds of the solution determined by IM, and only a few of vertices of the solution domain determined by CPM are inside the solution domain determined by IM. That is, CPM gives a smaller solution domain than IM. The fundamental factor that results in the overestimation of IM lies in neglecting the correlation between the uncertain-but-bounded parameters and simplifying these uncertainties to independent parameters. On the contrary, the correlation between the uncertainties is considered in CPM to solve the problem directly. Therefore, the vertex solution theorem of CPM can theoretically give the exact bounds of the structural static displacements.

7.6.2. *Ten-bar truss*

Next, we demonstrate the validity of the vertex solution theorem of CPM, by calculating the upper and lower bounds of the static displacements of the ten-bar truss shown in Figure 7.5, and the vertex solution theorem of IM is considered for comparison. In this example, two uncertain external loads P_1 and P_2 are acted on nodes 3 and 5, respectively. The cross-sectional areas of the 10 members are also uncertain, but the areas of the members ①–⑥ are identically equal and taken as A_1 and the others are identically equal and taken as A_2. Young's modulus of the element material is deterministic and taken as $E = 200$ GPa. The length of the horizontal or vertical members is $L = 1$ m.

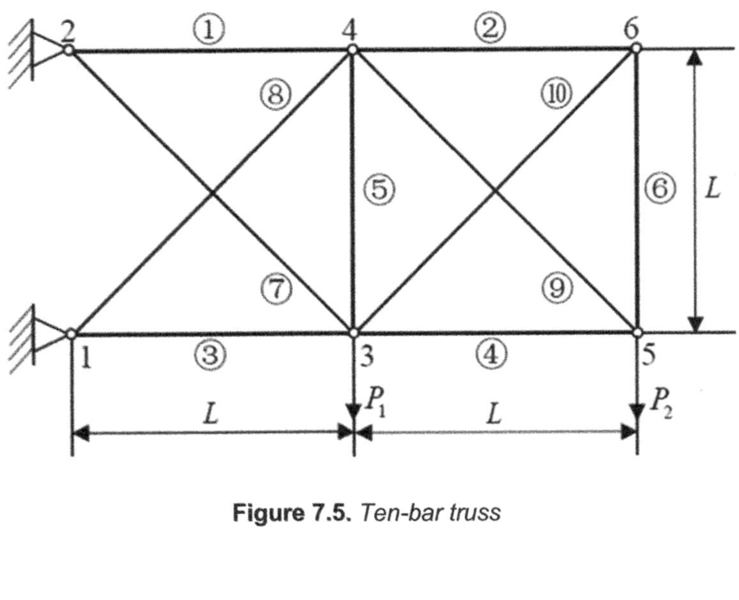

Figure 7.5. *Ten-bar truss*

The external load vector $\mathbf{P}=(P_1 \quad P_2)^{\mathrm{T}}$ and the cross-sectional area vector $\mathbf{A}=(A_1 \quad A_2)^{\mathrm{T}}$ are subject to the following constraint conditions, respectively:

$$\mathbf{P}\in\mathbf{P}^S(\boldsymbol{\alpha})=\left\{\mathbf{P}:\mathbf{P}=\sum_{p=1}^{4}\alpha_p\mathbf{P}_p,\ \sum_{p=1}^{4}\alpha_p=1,\ \alpha_p\geq 0,\ p=1,2,3,4\right\} \qquad [7.66]$$

and

$$\mathbf{A}\in\mathbf{A}^S(\boldsymbol{\beta})=\left\{\mathbf{A}:\mathbf{A}=\sum_{q=1}^{4}\beta_q\mathbf{A}_q,\ \sum_{q=1}^{4}\beta_q=1,\ \beta_q\geq 0,\ q=1,2,3,4\right\} \qquad [7.67]$$

where $\mathbf{P}^S$ and $\mathbf{A}^S$ denote the set combined by the known external load vertex vectors $\mathbf{P}_p\ (p=1,2,3,4)$ and the set combined by the known cross-sectional area vectors $\mathbf{A}_q\ (q=1,2,3,4)$, respectively, which are expressed as

$$\begin{aligned}&\mathbf{P}_1=\begin{pmatrix}200(1-\gamma)\mathrm{N}\\300(1-\gamma)\mathrm{N}\end{pmatrix},\quad \mathbf{P}_2=\begin{pmatrix}200(1+\gamma)\mathrm{N}\\300(1-\gamma)\mathrm{N}\end{pmatrix},\\&\mathbf{P}_3=\begin{pmatrix}200(1+\gamma)\mathrm{N}\\300(1+\gamma)\mathrm{N}\end{pmatrix},\quad \mathbf{P}_4=\begin{pmatrix}200(1-\gamma)\mathrm{N}\\300(1+\gamma)\mathrm{N}\end{pmatrix}\end{aligned} \qquad [7.68]$$

and

$$\begin{aligned}&\mathbf{A}_1=\begin{pmatrix}1\times10^{-4}\mathrm{m}^2\\1.2(1-\gamma)\times10^{-4}\mathrm{m}^2\end{pmatrix},\quad \mathbf{A}_2=\begin{pmatrix}(1+\gamma)\times10^{-4}\mathrm{m}^2\\1.2\times10^{-4}\mathrm{m}^2\end{pmatrix},\\&\mathbf{A}_3=\begin{pmatrix}1\times10^{-4}\mathrm{m}^2\\1.2(1+\gamma)\times10^{-4}\mathrm{m}^2\end{pmatrix},\quad \mathbf{A}_4=\begin{pmatrix}(1-\gamma)\times10^{-4}\mathrm{m}^2\\1.2\times10^{-4}\mathrm{m}^2\end{pmatrix}\end{aligned} \qquad [7.69]$$

where γ is the uncertain factor.

The convex polytopic models of the external loads P_1, P_2 and the cross-sectional areas A_1, A_2 are shown in Figure 7.6(a) and (b), respectively.

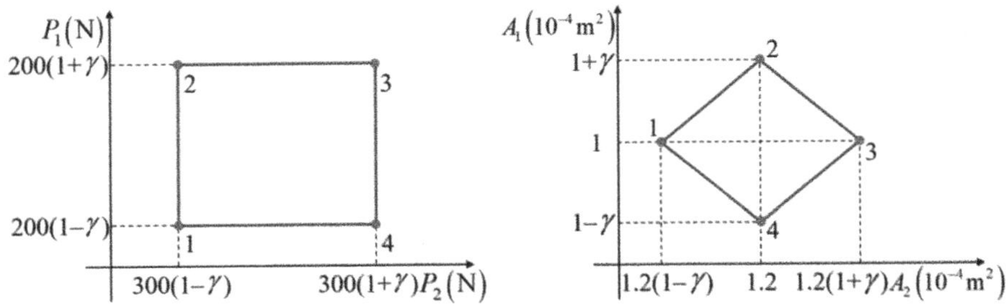

(a) External loads (b) Cross-sectional areas

Figure 7.6. *Convex polytopic models of the external loads and the cross-sectional areas of the ten-bar truss. For a color version of this figure, see www.iste.co.uk/challamel/mechanics3.zip*

For comparison, the upper and lower bounds of the vertical displacements u at nodes 3, 4, 5 and 6 when $\gamma = 5\%$, obtained by the vertex solution theorem based on CPM and IM, are shown in Table 7.1 and Figure 7.7. The distances of the upper and lower bounds are also included in Table 7.1.

u	CPM			IM		
	$u_{\min}$	$u_{\max}$	Δu	$u_{\min}$	$u_{\max}$	Δu
u_3	-4.710	-3.922	0.7882	-4.751	-3.889	0.8619
u_4	-1.803	-1.410	0.3933	-1.842	-1.376	0.4666
u_5	-12.58	-10.46	2.114	-12.68	-10.38	2.300
u_6	-6.963	-5.532	1.430	-7.059	-5.444	1.615

Table 7.1. *Bounds of the vertical displacements at nodes 3, 4, 5 and 6 of the ten-bar truss when* $\gamma = 5\%$ *(10^{-6} m)*

It can be observed from Table 7.1 and Figure 7.7 that the upper bounds of the vertical displacements obtained by CPM are smaller than those obtained by IM and the lower bounds of the vertical displacements obtained by CPM are larger than those obtained by IM. This suggests that CPM gives tighter bounds for the displacements than IM, which can also be certified by the distances of the upper and lower bounds listed in Table 7.1. The reason for this is that the overestimation of the bounds obtained by IM is revealed.

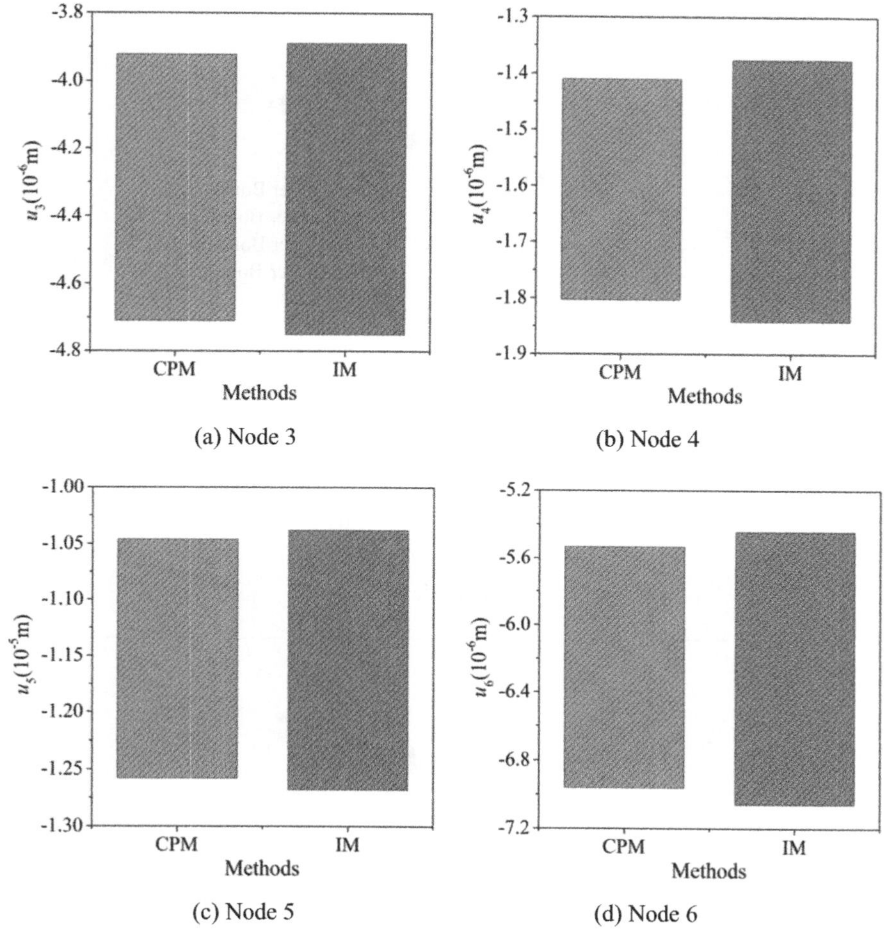

Figure 7.7. *Bounds of the vertical displacements at nodes 3, 4, 5 and 6 of the ten-bar truss when* $\gamma = 5\%$ *. For a color version of this figure, see www.iste.co.uk/challamel/mechanics3.zip*

To be more illustrative, we suppose that the uncertain factor γ varies in some region so that we can check that the static displacement bounds change monotonically with increasing uncertain factor γ. Comparison of the upper and lower bounds of the vertical displacements at nodes 5 and 6, corresponding to the uncertain factor γ ranging from 0 to 0.1 between the vertex solution theorem of CPM and IM, is shown in Figure 7.8.

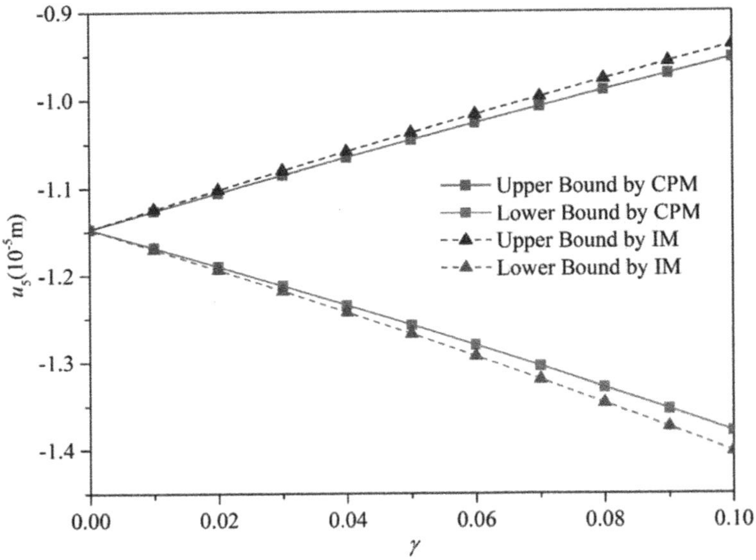

(a) Node 5

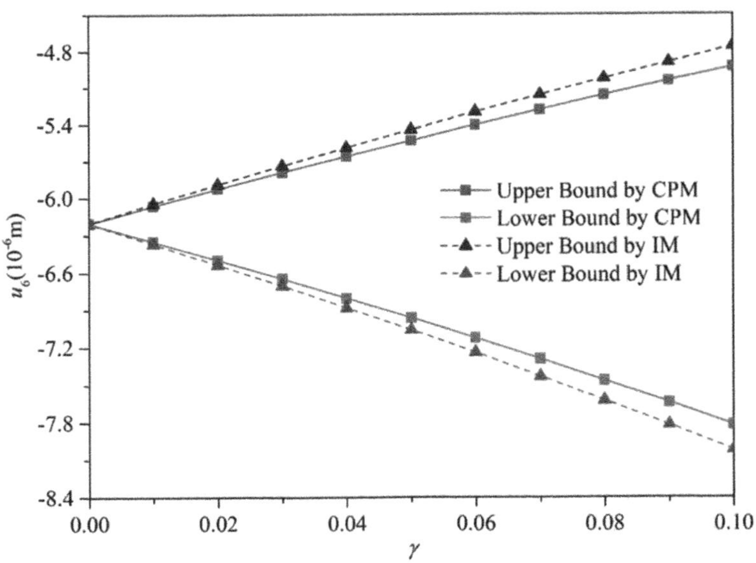

(b) Node 6

Figure 7.8. *Comparison of the upper and lower bounds on the vertical displacement at nodes 5 and 6 of the ten-bar truss, versus the uncertain factor γ. For a color version of this figure, see www.iste.co.uk/challamel/mechanics3.zip*

It can be seen from Figure 7.8 that as the uncertain factor γ increases, the upper bounds of the vertical displacements obtained by CPM and IM both increase and the lower bounds of the vertical displacements obtained by CPM and IM both decrease. It is worth noting that CPM yields narrower bounds of the displacements than IM. In other words, the upper bounds of the displacements obtained by CPM are smaller than those obtained by IM and the lower bounds of the displacements obtained by CPM are larger than those obtained by IM. For very low values of the uncertain factor γ, the upper and lower bounds obtained by IM are close to those obtained by CPM, respectively. With the increase of the uncertain factor γ, the overestimation of the bounds obtained by IM becomes serious and the accuracy of the results deteriorates. Under this circumstance, the table and the figures above reveal the fact that we can predict the exact bounds of the static displacements by the vertex solution theorem of CPM.

7.6.3. *Plane frame*

In this example, we consider a more complex structure, the plane frame with 202 nodes and 357 beam elements, as shown in Figure 7.9, to verify the effectiveness of the vertex solution theorem of CPM. The physical parameters of the plane frame are listed in Table 7.2. Due to errors in manufacturing, Young's modulus of the horizontal beams E_1 and the vertical beams E_2 are uncertain. The plane frame is subject to three uncertain external loads F_1, F_2 and F_3 acting on nodes 193, 192 and 198, respectively.

Young's modulus vector $\mathbf{E} = \left(E_1 \quad E_2\right)^{\mathrm{T}}$ and the external load vector $\mathbf{F} = \left(F_1 \quad F_2 \quad F_3\right)^{\mathrm{T}}$ are uncertain ranging inside the following convex polyhedrons, respectively

$$\mathbf{E} \in \mathbf{E}^S(\boldsymbol{\alpha}) = \left\{ \mathbf{E} : \mathbf{E} = \sum_{p=1}^{6} \alpha_p \mathbf{E}_p,\ \sum_{p=1}^{6} \alpha_p = 1,\ \alpha_p \geq 0,\ p = 1, 2, \cdots, 6 \right\} \qquad [7.70]$$

and

$$\mathbf{F} \in \mathbf{F}^S(\boldsymbol{\beta}) = \left\{ \mathbf{F} : \mathbf{F} = \sum_{q=1}^{8} \beta_q \mathbf{F}_q,\ \sum_{q=1}^{8} \beta_q = 1,\ \beta_q \geq 0,\ q = 1, 2, \cdots, 8 \right\} \qquad [7.71]$$

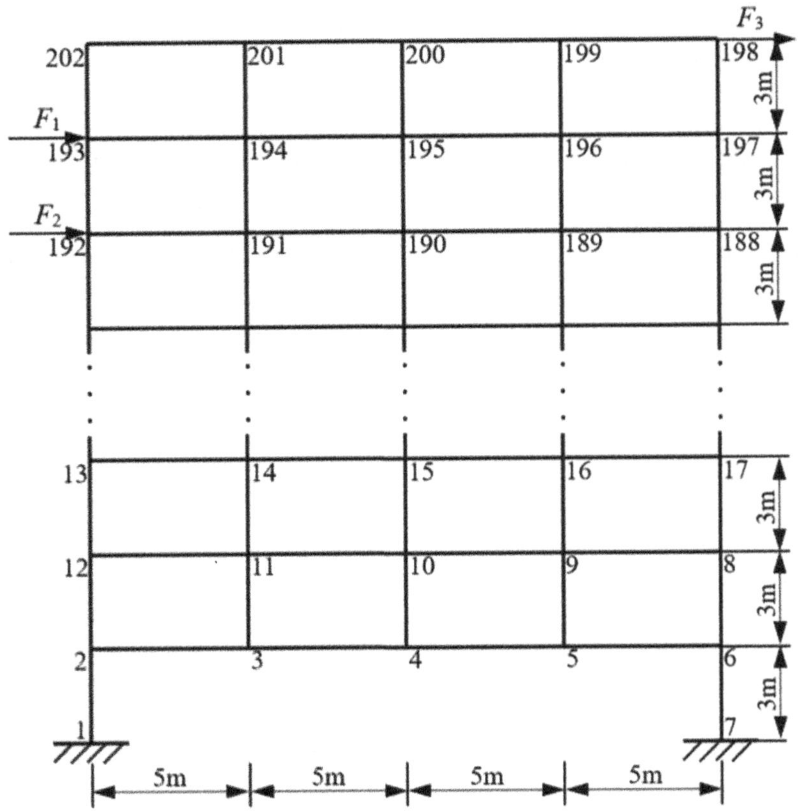

Figure 7.9. *Plane frame*

	Length	Height of the cross-section	Width of the cross-section
Horizontal beams	5	0.8	0.5
Vertical beams	3	0.6	0.5

Table 7.2. *Physical parameters of the plane frame (m)*

where $\mathbf{E}^S$ and $\mathbf{F}^S$ denote the set combined by the known Young's modulus vertex vectors $\mathbf{E}_p, p = 1, 2, \cdots, 6$, and the set combined by the known external load vectors $\mathbf{F}_q, q = 1, 2, \cdots, 8$, respectively. That is,

$$\mathbf{E}_1 = \begin{pmatrix} 210(1-0.5\gamma)\text{GPa} \\ 210(1-\gamma)\text{GPa} \end{pmatrix},\quad \mathbf{E}_2 = \begin{pmatrix} 210(1+0.5\gamma)\text{GPa} \\ 210(1-\gamma)\text{GPa} \end{pmatrix},$$
$$\mathbf{E}_3 = \begin{pmatrix} 210(1+\gamma)\text{GPa} \\ 210\text{GPa} \end{pmatrix},\quad \mathbf{E}_4 = \begin{pmatrix} 210(1+0.5\gamma)\text{GPa} \\ 210(1+\gamma)\text{GPa} \end{pmatrix}, \qquad [7.72]$$
$$\mathbf{E}_5 = \begin{pmatrix} 210(1-0.5\gamma)\text{GPa} \\ 210(1+\gamma)\text{GPa} \end{pmatrix},\quad \mathbf{E}_6 = \begin{pmatrix} 210(1-\gamma)\text{GPa} \\ 210\text{GPa} \end{pmatrix}$$

and

$$\mathbf{F}_1 = \begin{pmatrix} 10\text{kN} \\ 5(1-\gamma)\text{kN} \\ 10(1-\gamma)\text{kN} \end{pmatrix},\quad \mathbf{F}_2 = \begin{pmatrix} 10(1+\gamma)\text{kN} \\ 5(1-\gamma)\text{kN} \\ 10\text{kN} \end{pmatrix},\quad \mathbf{F}_3 = \begin{pmatrix} 10\text{kN} \\ 5(1-\gamma)\text{kN} \\ 10(1+\gamma)\text{kN} \end{pmatrix},$$
$$\mathbf{F}_4 = \begin{pmatrix} 10(1-\gamma)\text{kN} \\ 5(1-\gamma)\text{kN} \\ 10\text{kN} \end{pmatrix},\quad \mathbf{F}_5 = \begin{pmatrix} 10\text{kN} \\ 5(1+\gamma)\text{kN} \\ 10(1-\gamma)\text{kN} \end{pmatrix},\quad \mathbf{F}_6 = \begin{pmatrix} 10(1+\gamma)\text{kN} \\ 5(1+\gamma)\text{kN} \\ 10\text{kN} \end{pmatrix}, \qquad [7.73]$$
$$\mathbf{F}_7 = \begin{pmatrix} 10\text{kN} \\ 5(1+\gamma)\text{kN} \\ 10(1+\gamma)\text{kN} \end{pmatrix},\quad \mathbf{F}_8 = \begin{pmatrix} 10(1-\gamma)\text{kN} \\ 5(1+\gamma)\text{kN} \\ 10\text{kN} \end{pmatrix}$$

where γ is the uncertain factor.

The convex polytopic models of Young's moduli E_1, E_2 and the external loads F_1, F_2, F_3 are plotted in Figure 7.10(a) and (b), respectively.

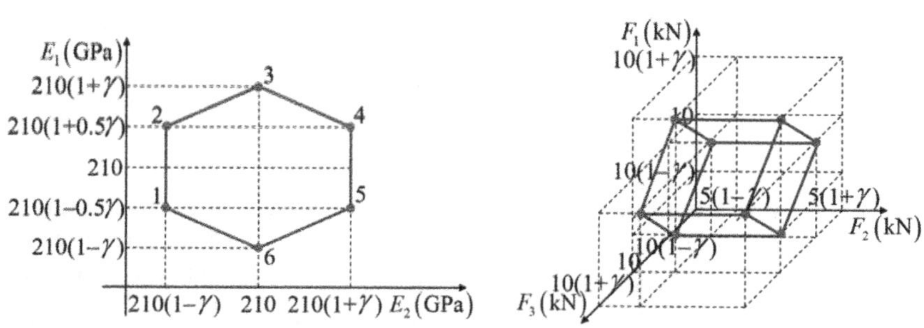

(a) Young's moduli (b) External loads

Figure 7.10. *Convex polytopic models of Young's moduli and the external loads of the plane frame. For a color version of this figure, see www.iste.co.uk/challamel/mechanics3.zip*

The upper and lower bounds of the horizontal displacements u at nodes 188, 192, 198 and 202, and the distances of the bounds when $\gamma = 3\%$, calculated by the vertex solution theorem based on CPM and IM, are shown in Table 7.3 and Figure 7.11.

From the numerical results in Table 7.3 and Figure 7.11, it can be seen that CPM yields tighter bounds for the displacements than IM, namely the upper bounds calculated by CPM are slightly smaller than those calculated by IM and the lower bounds calculated by CPM are slightly larger than those calculated by IM. Furthermore, from the distances of the bounds listed in Table 7.3, it can be found that the distances of IM are about 1.25 times as long as those of CPM.

In order to investigate the effect of the uncertain factor γ on the static displacement bounds, we consider the uncertain factor γ ranging from 0 to 0.1. The comparison of the upper and lower bounds of the horizontal displacements at nodes 188 and 198, versus the uncertain factor γ between the vertex solution theorem of CPM and IM, is shown in Figure 7.12.

u	CPM			IM		
	u_{min}	u_{max}	Δu	u_{min}	u_{max}	Δu
u_{188}	1.247	1.373	0.126	1.232	1.389	0.157
u_{192}	1.251	1.377	0.126	1.235	1.393	0.158
u_{198}	1.343	1.478	0.135	1.327	1.496	0.169
u_{202}	1.346	1.482	0.136	1.330	1.500	0.170

Table 7.3. *Bounds of the horizontal displacements at nodes 188, 192, 198 and 202 of the plane frame when* $\gamma = 3\%$ *(m)*

We can observe from Figure 7.12 that for very low values of the uncertain factor γ, the difference in accuracy of CPM and IM is small. With the increase of the uncertain factor γ, the upper bounds of the displacements obtained by IM increase more greatly than those obtained by CPM and the lower bounds of the displacements obtained by CPM decrease more slowly than those obtained by IM.

That is, the overestimation of the bounds calculated by IM becomes increasingly serious. As expected, it is demonstrated again that CPM gives tighter bounds for the displacements than IM. The good performance of the present method can be attributed to the fact that the vertex solution theorem of CPM can theoretically predict the exact bounds of the structural static displacements.

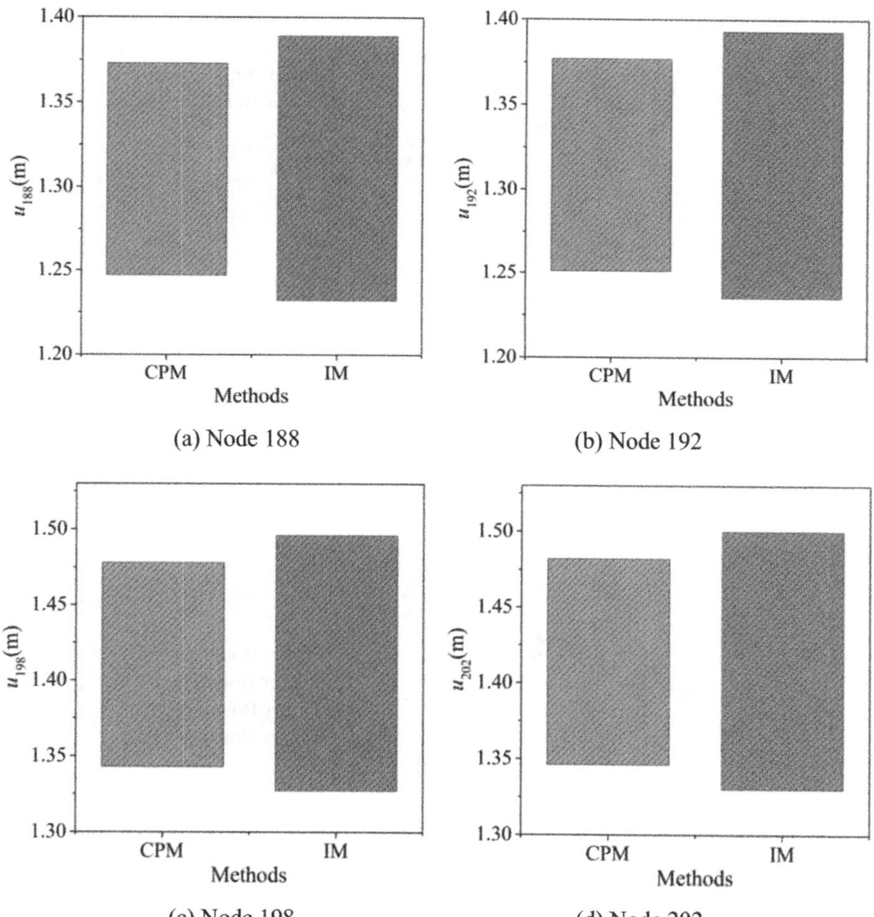

Figure 7.11. *Bounds of the horizontal displacements at nodes 188, 192, 198 and 202 of the plane frame when* $\gamma = 3\%$. *For a color version of this figure, see www.iste.co.uk/challamel/mechanics3.zip*

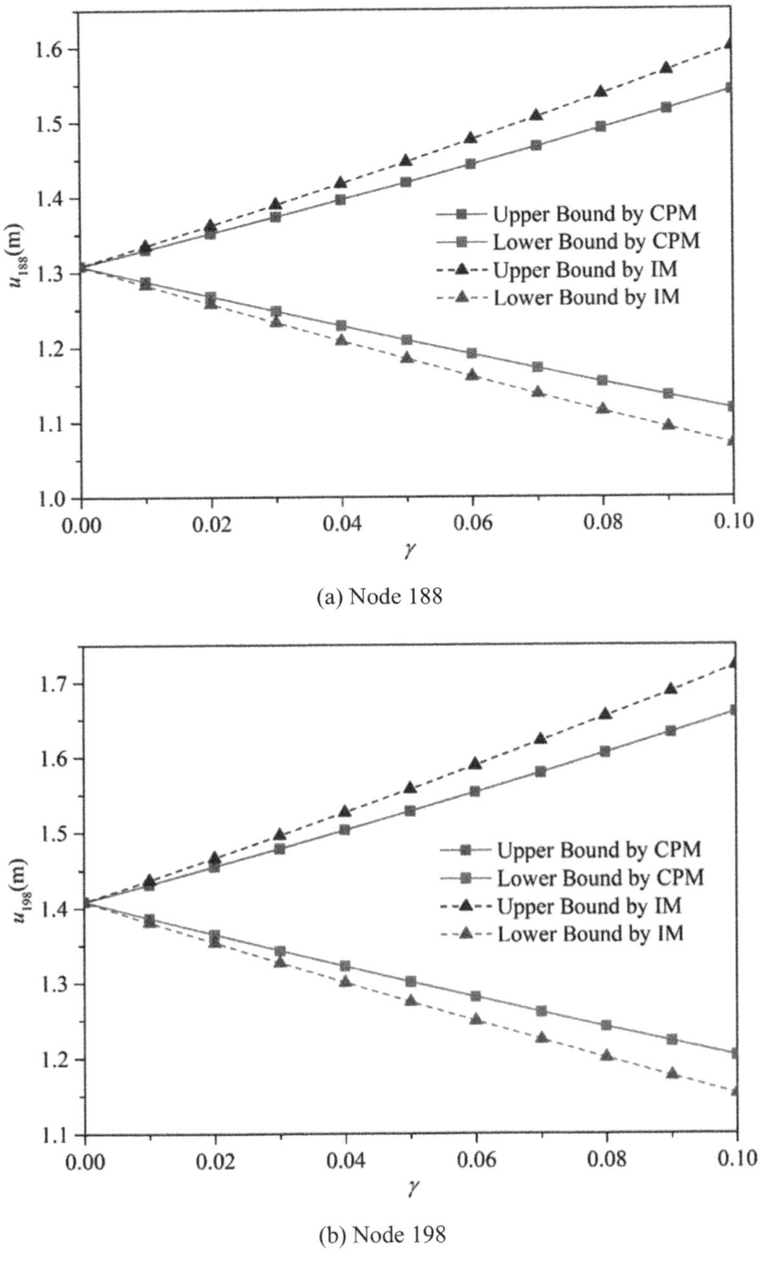

Figure 7.12. *Comparison of the upper and lower bounds on the horizontal displacement at nodes 188 and 198 of the plane frame, versus the uncertain factor γ. For a color version of this figure, see www.iste.co.uk/challamel/mechanics3.zip*

7.7. Conclusion

By virtue of the convex set theory, this chapter proposes a convex polytopic model for the static response of structures where the range of each uncertain-but-bounded parameter is treated as a closed and bounded convex polyhedron. Hence, the static response equation of structures can be expressed as a convex polytopic model. Its corresponding vertex solution theorem is deduced and validated when the convex polytopic model for the static response of structures is transformed into linear programming problems. As a result, the upper and lower bounds of the structural static displacements can only be easily and accurately obtained by calculating the vertices of the feasible regions of a series of linear programming subproblems, which are generated by combining the vertex matrices of the structural stiffness convex polyhedron and the vertex vectors of the external load convex polyhedron. Furthermore, in order to account for the validity and accuracy of the proposed method, three numerical examples are taken, including a two-step bar, a ten-bar truss and a plane frame. From the three numerical examples, we can observe that the vertex solution theorem of the convex polytopic model can give the exact bounds of the structural static displacements, overcoming the drawbacks of the interval model due to neglecting the correlation between the uncertain-but-bounded parameters. Therefore, the proposed method can play a significant role in static displacement bound estimation and become a powerful tool in static response analysis with uncertainties. Moreover, it can be highly applicable to the fields of engineering practice and will make great contributions.

7.8. Acknowledgments

The authors would like to thank the National Nature Science Foundation of the P. R. China (no. 11772026), the Defense Industrial Technology Development Program (nos. JCKY2016204B101 and JCKY2018601B001), the Beijing Municipal Science and Technology Commission via project (Z191100004619006) and the Beijing Advanced Discipline Center for Unmanned Aircraft System for their financial support. Furthermore, many thanks to Prof. Elishakoff for encouraging and supporting the authors in this research field.

7.9. References

Ben-Haim, Y. and Elishakoff, I. (1990). *Convex Models of Uncertainty in Applied Mechanics*. Elsevier, Amsterdam.

Dong, W. and Shah, H.C. (1987). Vertex method for computing functions of fuzzy variables. *Fuzzy Set. Syst.*, 24(1), 65–78.

Elishakoff, I., Eliseeff, P., Glegg, S. (1994). Convex modelling of material uncertainty in vibrations of a viscoelastic structure. *AIAA J.*, 32, 843–849.

Fujita, K. and Takewaki, I. (2011). An efficient methodology for robustness evaluation by advanced interval analysis using updated second-order Taylor series expansion. *Eng. Struct.*, 33(12), 3299–3310.

Ganzerli, S. and Pantelides, C.P. (2000). Optimum structural design via convex model superposition. *Comput. Struct.*, 74(6), 639–647.

Guo, X., Bai, W., Zhang, W.S. (2008). Extreme structural response analysis of truss structures under material uncertainty via linear mixed 0-1 programming. *Int. J. Numer. Methods Eng.*, 76(3), 253–277.

Horisberger, H.P. and Belanger, P.R. (1976). Regulators for linear, time invariant plants with uncertain parameters. *IEEE Trans. Automat. Control.*, 21(5), 705–708.

Jalivand-Nejad, A., Shafaei, R., Shahriari, H. (2016). Robust optimization under correlated polyhedral uncertainty set. *Comput. Ind. Eng.*, 92, 82–94.

Jiang, C., Han, X., Lu, G.Y., Liu, J., Zhang, Z., Bai, Y.C. (2011). Correlation analysis of non-probabilistic convex model and corresponding structural reliability technique. *Comput. Method Appl. Mech.*, 200(33–36), 2528–2546.

Jiang, C., Zhang, Z.G., Zhang, Q.F., Han, X., Xie, H.C., Liu, J. (2014). A new nonlinear interval programming method for uncertain problems with dependent interval variables. *Eur. J. Oper. Res.*, 238(1), 245–253.

Kanno, Y. and Takewaki, I. (2006). Confidence ellipsoids for static response of trusses with load and structural uncertainties. *Comput. Method Appl. Mech.*, 196(1–3), 393–403.

Köylüoğlu, H.U., Cakmak, A.S., Nielsen, S.R.K. (1995). Interval algebra to deal with pattern loading and structural uncertainties. *J. Eng. Mech.*, 121(11), 1149–1157.

Li, Q., Qiu, Z.P., Zhang, X.D. (2017). Static response analysis of structures with interval parameters using the second-order Taylor series expansion and the DCA for QB. *Acta Mech. Sin.*, 31(6), 845–854.

Liu, Y.S., Wang, X.J., Wang, L., Lv, Z. (2019). A Bayesian collocation method for static analysis of structures with unknown-but-bounded uncertainties. *Comput. Method Appl. Mech.*, 346, 727–745.

Luo, Z.X., Wang, X.J., Liu, D.L. (2020). Prediction on the static response of structures with large-scale uncertain-but-bounded parameters based on the adjoint sensitivity analysis. *Struct. Multidiscip. O.*, 61(1), 123–139.

McWilliams, S. (2001). Anti-optimization of uncertain structures using interval analysis. *Comput. Struct.*, 79(4), 421–430.

Mo, L.P. (2011). Robust stabilization for multi-input polytopic nonlinear systems. *J. Syst. Sci. Complex.*, 24(1), 93–104.

Moore, R.E., Kearfott, R.B., Cloud, M.J. (2009). *Introduction to Interval Analysis*. SIAM, Philadelphia.

Muhanna, R.L. and Mullen, R.L. (1999). Bounds of structural response for all possible loading combinations. *J. Struct. Eng.*, 125(1), 98–106.

Muhanna, R.L., Zhang, H., Mullen, R.L. (2007). Interval finite elements as a basis for generalized models of uncertainty in engineering mechanics. *Reliab. Comput.*, 13(2), 173–194.

Oettli, W. and Prager, W. (1964). Compatibility of approximate solution of linear equations with given error bounds for coefficients and right-hand sides. *Numer. Math.*, 6(1), 405–409.

de Oliveira, M.C., Geromel, J.C., Bernussou, J. (2002). Extended H_2 and H_∞ norm characterizations and controller parametrizations for discrete-time systems. *Int. J. Control*, 75(9), 666–679.

Pantelides, C.P. and Ganzerli, S. (2001). Comparison of fuzzy set and convex model theories in structural design. *Mech. Syst. Signal Pr.*, 15(3), 499–511.

Qiu, Z.P. (2003). Comparison of static response of structures using convex models and interval analysis method. *Int. J. Numer. Methods Eng.*, 56(12), 1735–1753.

Qiu, Z.P. and Elishakoff, I. (1998). Antioptimization of structures with large uncertain-but-non-random parameters via interval analysis. *Comput. Method Appl. Mech.*, 152(3), 361–372.

Qiu, Z.P. and Jiang, N. (2021). An ellipsoidal Newton's iteration method of nonlinear structural systems with uncertain-but-bounded parameters. *Comput. Method Appl. Mech.*, 373, 113501.

Qiu, Z.P. and Liu, D.L. (2020). Safety margin analysis of buckling for structures with unknown but bounded uncertainties. *Appl. Math. Comput.*, 367, 124759.

Qiu, Z.P. and Lv, Z. (2017). The vertex solution theorem and its coupled framework for static analysis of structures with interval parameters. *Int. J. Numer. Methods Eng.*, 112(7), 711–736.

Qiu, Z.P. and Lyu, Z. (2020). Vertex combination approach for uncertainty propagation analysis in spacecraft structural system with complex eigenvalue. *Acta Astronaut.*, 171, 106–117.

Qiu, Z.P., Xia, Y.Y., Yang, J.L. (2007). The static displacement and the stress analysis of structures with bounded uncertainties using the vertex solution theorem. *Comput. Method Appl. Mech.*, 196(49–52), 4965–4984.

Rao, S.S. (1984). *Optimization: Theory and Applications*. Halsted Press, New York.

Rao, S.S. and Chen, L. (1998). Numerical solution of fuzzy linear equations in engineering analysis. *Int. J. Numer. Methods Eng.*, 43(3), 391–408.

Sriramula, S. and Chryssanthopoulos, M.K. (2013). An experimental characterisation of spatial variability in GFRP composite panels. *Struct. Saf.*, 42, 1–11.

Wang, L. and Liu, Y.R. (2020). A novel method of distributed dynamic load identification for aircraft structure considering multi-source uncertainties. *Struct. Multidiscip. Optim.*, 61(5), 1929–1952.

Wang, L. and Xiong, C. (2019). A novel methodology of sequential optimization and non-probabilistic time-dependent reliability analysis for multidisciplinary systems. *Aerosp. Sci. Technol.*, 94, 105389.

Wang, X.J., Elishakoff, I., Qiu, Z.P. (2008). Experimental data have to decide which of the nonprobabilistic uncertainty descriptions – convex modeling or interval analysis – to utilize. *J. Appl. Mech.*, 75(4), 1–8.

Xu, M.H. and Qiu, Z.P. (2014). A dimension-wise method for the static analysis of structures with interval parameters. *Sci. China Phys. Mech.*, 57(10), 1934–1945.

Zhou, Y.T., Jiang, C., Han, X. (2006). Interval and subinterval analysis methods of the structural analysis and their error estimations. *Int. J. Comp. Methods*, 3(2), 229–244.

8

On the Interval Frequency Response of Cracked Beams with Uncertain Damage

The present contribution addresses the frequency response of damaged beams in presence of cracks characterized by interval uncertainty, setting a correspondence between two alternative models for describing the cracks. Under the assumption of a fully open crack during vibrations, the damage is represented by the widely used "local flexibility model" following both the "*finite element*" and "*continuous*" approaches. In the first category, the stiffness reduction of the cracked member is conveniently evaluated by energy balance in terms of stress intensity factors and the depth of the crack is assumed as an interval parameter. The second approach models the crack as a linearly elastic rotational spring with stiffness taken as an *uncertain-but-bounded* parameter. A direct comparison between the responses adopting the finite element and continuous models allows us to fix a correspondence between the interval uncertainty in the "crack depth" and in its "spring stiffness" counterpart. Lower and upper bounds of the frequency response are evaluated following two alternative procedures. The first one, developed for discrete damaged beams, takes full advantage of the approximate explicit expression provided for the interval frequency response function; the second one, introduced to handle continuous damaged beams, is based on a sensitivity-based method.

Numerical applications on a multi-cracked beam show the effectiveness of both the proposed procedures returning very accurate results in terms of deflection function bounds and confirm the correspondence in the interval modeling of the crack depth as well as the spring stiffness.

Chapter written by Roberta SANTORO.

8.1. Introduction

Over the last years, several studies have been conducted on the dynamics of cracked beams. The presence of cracks is one of the most important causes of accidents and failures in structures. Therefore, investigations into vibrations of cracked beam-like structures are of great interest for the scientific community.

Under the assumption of an "always open-crack model" during vibrations, the simplest analytical model for the computation of dynamic response of a damaged Euler–Bernoulli beam is to treat the cracked region as a local flexibility. Based on this, the theoretical modeling techniques of cracked beams can be grouped into the "*finite element models*" (*FE*) and the "*continuous models*".

In the first category, the presence of a crack is modeled by an *ad hoc* finite element with reduced flexural flexibility conveniently evaluated by energy balance in terms of stress intensity factors (Qian *et al.* 1990; Ruotolo *et al.* 1996).

For the second category, based on the common assumption of the crack effect as a concentrated flexural flexibility (e.g. Caddemi and Caliò (2009); Caddemi and Morassi (2013) and references therein), a crack is modeled by a rotational spring connecting two adjacent beam cross-sections at the crack location (Chaudhari *et al.* 2000; Zhao *et al.* 2016), with the spring stiffness related to the size of the crack (Rizos *et al.* 1990; Ostachowicz and Krawczuk 1991; Chondros *et al.* 1998; Fernandez-Saez *et al.* 1999; Bilello 2001).

Usually, in the deterministic fracture mechanics analysis, the crack size is assumed to be known; however, it possesses a considerable amount of uncertainty, and it is essential to take this into account, regardless of the modeling technique adopted. The unavoidable uncertainty can be described alternatively with probabilistic or non-probabilistic approaches. The probabilistic approach requires a lot of data to define the probability density function (PDF) of the uncertain parameters. If available information on uncertain parameters is fragmentary or incomplete, then non-probabilistic approaches (Elishakoff 1995, 2010) can be alternatively applied. In particular, the interval model is today the most used analytical tool and is based on the interval arithmetic first introduced by Moore (1996) and subsequent detailed studies on this subject (Muhanna and Mullen 2001; Moens and Vandepitte 2005; Elishakoff and Miglis 2012; Muscolino and Sofi 2012). Specifically, in the framework of interval analysis, the generic parameter is represented by an "interval" or equivalent labeled as an "uncertain-but-bounded" variable considering its variability within fixed ranges defined by lower and upper bounds.

Some previous studies have already considered the issue concerning the variation of the structural response due to the uncertainty in the crack parameters. In particular, attention has been focused on the dynamic response of damaged beams in the presence of a single crack or multiple cracks with uncertain crack depth described as a random variable (Cacciola and Muscolino 2002) or an interval variable (Muscolino and Santoro 2017a, 2017b, 2019; Santoro and Muscolino 2019) adopting in both cases the finite element model. The interval static response considering the most general case of uncertainty in size and position of the cracks has been recently evaluated resorting to a continuous model for the damaged beam (Santoro *et al.* 2020).

The aim of this chapter is to provide the dynamic response in the frequency domain of a multi-cracked beam where the crack depth is assumed as an interval variable and the crack is described following both the finite element model and the continuous one.

First, a direct comparison between the responses in terms of deflection function adopting the finite element and continuous approaches allows us to fix a correspondence between the interval uncertainty characterizing the crack depth and its counterpart in the spring stiffness.

Once this correspondence is defined, the interval frequency response of a multi-cracked beam is evaluated by adopting both models (the FE and the continuous models) and applying two alternative procedures. The first one, developed for discretized damaged beams, takes full advantage of the approximate explicit expression introduced for the interval frequency response function (Muscolino *et al.* 2014) and provides an extension on results evaluated for a single-cracked beam (Muscolino and Santoro 2017b); the second one, recently introduced to handle continuous multi-cracked beams in a static setting, is based on a sensitivity-based method (Santoro *et al.* 2020; Santoro and Failla 2021).

Both of the proposed procedures overcome the main shortcoming of the most common interval analysis methods based on the first-order Taylor expansion (Qiu and Wang 2003) and the interval perturbation (Qiu and Wang 2005; Santoro and Muscolino 2019). Despite their simplicity, the effectiveness of the mentioned methods is limited to small fluctuations of the interval parameters (Santoro and Muscolino 2019). Moreover, among the interval analysis approaches, it is worth mentioning the so-called *vertex method*, which represents the reference combinatorial solution providing exact response bounds under the assumption of monotonicity (Dong and Shah 1987; Moens and Vandepitte 2005). However, its computational effort increases exponentially with the increasing number of the uncertain parameters considered.

Numerical applications confirm the correspondence in the interval modeling of the crack depth as well as the spring stiffness, and show the effectiveness of both the proposed procedures, returning very accurate results in terms of the deflection function bounds regardless of the level of uncertainty.

8.2. Crack modeling for damaged beams

8.2.1. *Finite element crack model*

The analytical model adopted to treat discretized damaged beams with a transverse on-edge non-propagating crack is based on the finite element model proposed by Qian *et al.* (1990) and Ruotolo *et al.* (1996).

According to the Saint-Venant principle, the damage influences the stress field in the crack proximity. Such a perturbation is relevant, especially when the crack is open and determines a local reduction of the flexural rigidity in the region adjacent to the damage. Based on this observation, only the finite element that contains a central crack is subjected to a stiffness modification (see Figure 8.1).

The generic component $d_{ik}^{(0)}$ of the compliance (or flexibility) matrix $\mathbf{D}_{\mathrm{e}}^{(0)}$ related to the undamaged element and the generic term $d_{ik}^{(1)}$ of the additional flexibility matrix $\mathbf{D}_{\mathrm{e}}^{(1)}$ accounting for the crack can be derived, respectively, as

$$d_{ik}^{(0)} = \frac{\partial^2 W^{(0)}}{\partial P_i \partial P_k}, \quad d_{ik}^{(1)} = \frac{\partial^2 W^{(1)}}{\partial P_i \partial P_k}; \quad i,k = 1,2; \quad P_1 = S, \quad P_2 = M \qquad [8.1]$$

with $W^{(0)}$ being the strain energy for the intact element, $W^{(1)}$ the additional strain energy due to the crack and S and M the shear and bending internal forces at the right node of the element, respectively. Specifically, the evaluation of $W^{(1)}$ has been studied in fracture mechanics. In detail, $W^{(0)}$ and $W^{(1)}$ given in equation [8.1] are, respectively, expressed by:

$$W^{(0)} = \frac{1}{2EI}\left(M^2 \ell + \frac{S^2 \ell^3}{3} + MS\ell^2 \right); \; W^{(1)} = b\int_0^a \frac{\left(K_{IM} + K_{IS}\right)^2 + K_{IIS}^2}{E} da \qquad [8.2]$$

where E is the Young modulus, I is the moment of inertia and ℓ is the length of the finite element. The additional energy $W^{(1)}$ in its simplified form (taking into account only bending) as reported in equation [8.2] is written for a prismatic beam having width b and thickness h; "a" denotes the crack depth and K_{IM}, K_{IS}, K_{IIS} are the stress intensity factors for opening-type and sliding-type cracks, due to M and S, respectively (for detailed expressions, see, for example, Qian *et al.* (1990) and Ruotolo *et al.* (1996)).

The undamaged parts of the beam are modeled by Euler-type finite elements with two nodes and two degrees of freedom (transverse displacement and rotation) at each node.

Finally, the total flexibility matrix for the generic element "e" with an open crack is provided by:

$$\mathbf{D}_{\mathrm{e}} = \mathbf{D}_{\mathrm{e}}^{(0)} + \mathbf{D}_{\mathrm{e}}^{(1)} \qquad [8.3]$$

and by the principle of virtual work, the stiffness matrices of the undamaged and cracked elements take, respectively, the following form:

$$\mathbf{K}_{\mathrm{e}} = \mathbf{T}\mathbf{D}_{\mathrm{e}}^{(0)-1}\mathbf{T}^{T}; \quad \mathbf{K}_{\mathrm{c,e}} = \mathbf{T}\mathbf{D}_{\mathrm{e}}^{-1}\mathbf{T}^{T}; \quad \mathbf{T}^{T} = \begin{bmatrix} -1 & -\ell & 1 & 0 \\ 0 & -1 & 0 & 1 \end{bmatrix} \qquad [8.4]$$

where the apex T means transpose matrix.

Once the stiffness matrices of the undamaged and cracked elements for the beam discretized in N_e finite elements are defined, the global stiffness matrix $\mathbf{K}(a)$ of order $n \times n$ with $n = 2N_{\mathrm{e}}$ can be straightforwardly evaluated following the classical assembly rules.

The procedure presented to build the flexibility matrix for a cracked element (see equation [8.3]) can be repeated to consider the simultaneous presence of multiple cracks with depths $a_j = \hat{a}_j h$, with $j = 1,...,N_c$, N_c the number of cracks and $\hat{a}_j$ the dimensionless crack depth.

8.2.2. *Continuous crack model*

In the continuous beam model, the presence of an open crack results in a discontinuity in the slope of the beam deflection. Therefore, based on this consideration and according to a well-established approach in one-dimensional beam theory, the crack is used to be modeled as a massless rotational spring of stiffness "κ" connecting two intact segments (see Figure 8.1).

Such discontinuity is proportional to the bending moment transmitted through the cracked section. It follows that the relative rotation between the adjacent cross-sections denoted as $\Delta\Theta_j$ representing the discontinuity at the jth crack location $x = x_j$ is expressed by:

$$\Delta\Theta_j(\kappa_j) = \kappa_j^{-1} M(x_j) \qquad [8.5]$$

with $M(x_j)$ being the bending moment at $x = x_j$.

In the following, the spring stiffness κ_j for the generic jth spring corresponding to the jth crack is expressed in terms of the beam mechanical properties as:

$$\kappa_j = \hat{\kappa}_j \, EI/L \qquad [8.6]$$

where E and I, as previously introduced, represent the Young modulus and the moment of inertia, respectively, L is the total length of the beam and $\hat{\kappa}_j$ is the dimensionless spring stiffness.

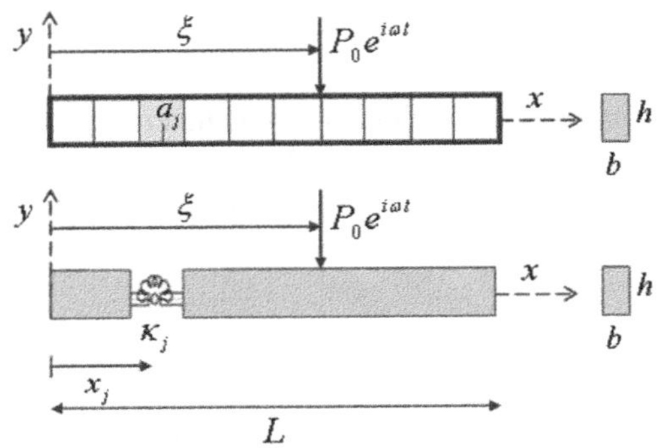

Figure 8.1. *Damaged beam: finite element model and continuous model. For a color version of this figure, see www.iste.co.uk/challamel/mechanics3.zip*

8.3. Statement of the problem

The interval model assumed for the crack depth is introduced in section 8.3.1. Next, the equations governing the dynamic response of a multi-cracked beam are reported in section 8.3.2 for both the discretized and continuous beam models. The correspondence between the uncertain crack depth and the uncertain spring stiffness representing the crack in the continuous model is described in section 8.3.3.

8.3.1. *Interval model for the uncertain crack depth*

In engineering applications, the crack parameters may be affected by unavoidable uncertainty, and it is essential to take account of it. Generally, uncertainties may affect the size as well as the position of the crack. In this chapter, all the analyses are conducted considering the uncertainty only in the depth a_j of the generic *j*th crack.

Following a non-probabilistic approach, the underlying idea is to represent the generic uncertain parameter a_j as an *interval* variable varying between a range specified by its lower and upper bounds. Equivalently, an interval parameter will be referred to as an *uncertain-but-bounded* variable.

Based on the interval analysis (Moore 1990), an interval variable α_j^I is defined as a bounded real number if $\alpha_j^I \triangleq \left[\underline{\alpha}_j, \overline{\alpha}_j\right] \in \mathbb{IR}$, where the apex *I* means interval variable, $\underline{\alpha}_j$ and $\overline{\alpha}_j$ denote the *lower bound* (LB) and the *upper bound* (UB) of α_j^I, respectively, and $\mathbb{IR}$ is the set of all closed intervals of real numbers. By introducing the central value $\alpha_{0,j}$ and the deviation amplitude $\Delta\alpha_j$, the interval variable α_j^I can be alternatively represented in the so-called *affine form* as

$$\alpha_j^I = \alpha_{0,j} + \Delta\alpha_j\, e_j^I \qquad [8.7]$$

with

$$\alpha_{0,j} = \frac{\underline{\alpha}_j + \overline{\alpha}_j}{2};\ \Delta\alpha_j = \frac{\overline{\alpha}_j - \underline{\alpha}_j}{2};\ e_j^I \triangleq [-1,1] \qquad [8.8]$$

where $e_j^I \triangleq [-1,1]$ represents the so-called extra-unitary interval associated with the *j*th variable.

In view of equation [8.7] and equation [8.8] and considering a symmetric fluctuation of the uncertainty around the nominal value (corresponding to $\alpha_{0,j} = 0$ in equation [8.8]), the uncertain crack depth can therefore be expressed by:

$$a_j^I = a_{0,j}(1+\alpha_j^I) = a_{0,j}(1+\Delta\alpha_j\, e_{a,j}^I) \qquad [8.9]$$

where $a_{0,j}$ represents the deterministic nominal value of the depth for the *j*th crack (no uncertainty is considered in the crack depth) and the fluctuation $\alpha_j^I = \Delta\alpha_j\, e_{a,j}^I$

provides the considered level of uncertainty. For a multi-cracked beam, the parameters $a_j \in a_j^I = \left[\underline{a}_j, \overline{a}_j\right]$, for j = 1,2...N_c (N_c = number of cracks), can be collected in the vector $\mathbf{a} \in \mathbf{a}^I = [\underline{\mathbf{a}}, \overline{\mathbf{a}}]$ expressed by

$$\mathbf{a}^I = \mathbf{a}_0(1+\boldsymbol{\alpha}^I) \qquad [8.10]$$

where, in turn, the vector $\boldsymbol{\alpha} \in \boldsymbol{\alpha}^I = [\underline{\boldsymbol{\alpha}}, \overline{\boldsymbol{\alpha}}]$ represents the dimensionless fluctuations of the uncertain crack depths around their nominal values collected in the vector $\mathbf{a}_0$.

For the linearly elastic rotational spring model, the uncertainty in the crack depth corresponds to assume the uncertainty in the stiffness κ_j of the jth spring.

Therefore, based on the interval approach, the spring stiffness κ_j^I will be an interval variable too, defined by its lower $\underline{\kappa}_j$ and upper $\overline{\kappa}_j$ bounds.

The aim of sections 8.3.2 and 8.3.3 is to define a correspondence between the interval model for the uncertain-but-bounded crack depth, as reported in equation [8.9], and the uncertain-but-bounded spring stiffness κ_j^I based on a direct comparison between the responses of damaged discretized and continuous beams.

8.3.2. *Governing equations of damaged beams*

Let us first introduce in the frequency domain the equation of motion for a damaged beam discretized in finite elements as well as for a damaged beam represented via its continuous model.

First, the equation of motion of a quiescent multi-cracked beam discretized by N_e finite elements subjected to an external deterministic excitation $\mathbf{f}(t)$ and neglecting for simplicity the damping effect can be written as:

$$\mathbf{M}\,\ddot{\mathbf{v}}(\mathbf{a}^I,t)+\mathbf{K}(\mathbf{a}^I)\,\mathbf{v}(\mathbf{a}^I,t)=\mathbf{f}(t) \qquad [8.11]$$

where $\mathbf{a}^I$ is the $N_c \times 1$ vector collecting the uncertain crack depths (see equation [8.10]), $\mathbf{M}$ is the $n \times n$ deterministic mass matrix, $\mathbf{K}(\mathbf{a}^I)$ is the $n \times n$ stiffness matrix and $\mathbf{f}(t)$ is the deterministic load vector of order $n \times 1$; $\mathbf{v}(\mathbf{a}^I,t)$ is the $n \times 1$ vector of nodal uncertain-but-bounded displacements and a dot over a variable denotes differentiation with respect to time t.

By performing the Fourier transform of both sides of equation [8.11], the following set of algebraic frequency-dependent equations governing the response in the frequency domain is obtained as

$$\mathbf{V}(\mathbf{a}^I,\omega)=\mathbf{H}(\mathbf{a}^I,\omega)\mathbf{F}(\omega) \qquad [8.12]$$

with

$$\mathbf{H}^I(\omega)=\mathbf{H}(\mathbf{a}^I,\omega)=\left[-\omega^2\mathbf{M}+\mathbf{K}(\mathbf{a}^I)\right]^{-1} \qquad [8.13]$$

where, in equations [8.12] and [8.13], $\mathbf{H}(\mathbf{a}^I,\omega)$ is the *frequency response function (FRF) matrix* (also referred to as the *transfer function matrix*) while $\mathbf{V}(\mathbf{a}^I,\omega)$ and $\mathbf{F}(\omega)$ are the vectors collecting the Fourier transforms of $\mathbf{v}(\mathbf{a}^I,t)$ and $\mathbf{f}(t)$, respectively. In the following explanation as well as in the numerical applications, the attention will be focused on a specific load condition, namely a transverse harmonically varying concentrated force acting at an arbitrary position $x=\xi$, as shown in Figure 8.1.

Let us now consider the counterpart of equation [8.12] for the continuous damaged beam model. Let $v=V(x,\omega)e^{i\omega t}$ be the steady-state flexural deflection response to a transverse harmonically varying point force $P_0e^{i\omega t}$ at fixed abscissa $x=\xi$ (see Figure 8.1). Using the theory of generalized functions (Yavari and Sarkani 2001; Falsone 2002), the steady-state motion equation of a uniform isotropic multi-cracked Euler–Bernoulli beam takes the following form:

$$EI\frac{\tilde{\mathrm{d}}^4V(x,\omega,\boldsymbol{\kappa}^I)}{\mathrm{d}x^4}-\sum_{j=1}^{N_c}EI\Delta\Theta_j(\kappa_j^I)\delta^2(x-x_j)-P_0\delta(x-\xi)-m\omega^2V(x,\omega,\boldsymbol{\kappa}^I)=0 \qquad [8.14]$$

where the "tilde" symbol denotes generalized derivative, $\delta(\bullet)$ is the Dirac delta function and $\delta^2(\bullet)$ is the formal second derivative of the Dirac delta. Moreover, in equation [8.14], m is the mass per unit length, $\Delta\Theta_j(\kappa_j^I)$ is the relative rotation between adjacent cross-sections accounting for the cracks as defined in equation [8.5] and the vector $\boldsymbol{\kappa}^I$ collects the uncertain-but-bounded stiffness κ_j^I of the N_c springs.

8.3.3. *Finite element model versus continuous model*

In this section, we focus on the correspondence between the values of the crack depth "a_j" and the spring stiffness "κ_j" starting from the several equivalent massless rotational spring models introduced in scientific literature to properly characterize the mechanical properties of the damaged sections (Zhao *et al.* 2016). Then, a direct comparison between the finite element model and the continuous one for a damaged beam allows us to find a similar relationship extended to the case of interval uncertain damage. The analysis conducted on a single-cracked beam can be easily extended to a multi-cracked beam. In the following, the subscript "*j*" referring to the *j*th crack as well as the *j*th spring is omitted for simplicity. Moreover, we first refer to the nominal value a_0 for the crack depth and the corresponding nominal value for the spring stiffness denoted by κ_0.

Among the spring models employed to describe the mechanical behavior of a cracked cross-section, some of them can be uniformly expressed by the following relationship:

$$\kappa_0 = \frac{EI}{h}\frac{1}{C(\hat{a}_0)} \qquad [8.15]$$

which connects the rotational spring stiffness κ_0 with the ratio $\hat{a}_0 = a_0/h$ between the crack depth a_0 and the height h of the beam cross-section. In equation [8.15], $C(\hat{a}_0)$ is the dimensionless local compliance function whose expression differs depending on the spring model. In the following, the closed-form expressions of $C(\hat{a}_0)$ are reported for the five most popular spring models.

Based on the results in terms of strain energy density developed in the fracture mechanics, Rizos *et al.* (1990) proposed the following polynomial expression:

$$\begin{aligned} C(\hat{a}_0) = 5.346\hat{a}_0^2 \big(&1.86 - 3.95\hat{a}_0 + 16.375\hat{a}_0^2 - 37.226\hat{a}_0^3 + 76.81\hat{a}_0^4 + \\ &-126.9\hat{a}_0^5 + 172\hat{a}_0^6 - 143.97\hat{a}_0^7 + 66.56\hat{a}_0^8\big) \end{aligned} \qquad [8.16]$$

Following a similar approach, different expressions of $C(\hat{a}_0)$ were provided by Ostachowicz *et al.* (1991) in the form:

$$\begin{aligned} C(\hat{a}_0) = 6\pi\hat{a}_0^2 \big(&0.6384 - 1.035\hat{a}_0 + 3.7201\hat{a}_0^2 - 5.1773\hat{a}_0^3 + \\ &+7.553\hat{a}_0^4 - 7.332\hat{a}_0^5 + 2.4909\hat{a}_0^6\big) \end{aligned} \qquad [8.17]$$

and by Fernandez-Saez *et al.* (1999) in the relation given as:

$$C(\hat{a}_0) = 2\left(\frac{\hat{a}_0}{1-\hat{a}_0}\right)^2 \left(5.93 - 19.69\hat{a}_0 + 37.14\hat{a}_0^{\,2} - 35.84\hat{a}_0^{\,3} + 13.12\hat{a}_0^{\,4}\right) \quad [8.18]$$

Other expressions were obtained by the so-called approach of lumped flexibility. In this framework, the approximate expression evaluated by Chondros *et al.* (1998) is:

$$\begin{aligned} C(\hat{a}_0) = 6\pi\left(1-\nu^2\right)\hat{a}_0^{\,2}\big(&0.6272 - 1.04533\hat{a}_0 + 4.5948\hat{a}_0^{\,2} - 9.9736\hat{a}_0^{\,3} + \\ &20.2948\hat{a}_0^{\,4} - 33.0351\hat{a}_0^{\,5} + 47.1063\hat{a}_0^{\,6} - 40.7556\hat{a}_0^{\,7} + 19.6\hat{a}_0^{\,8}\big) \end{aligned} \quad [8.19]$$

and the alternative simplest model suggested by Bilello (2001) is expressed as:

$$C(\hat{a}_0) = \frac{\hat{a}_0(2-h')}{0.9(1-\hat{a}_0)^2} \quad [8.20]$$

At this step, let us compare the response of the damaged discretized beam with that obtained by considering the continuous model aiming to provide a relationship similar to that reported in equation [8.15] relating the spring stiffness κ_0 with the crack depth a_0, which explicitly appears in the FE approach (see equation [8.2]).

Based on the consideration that the presence of a crack affects the stiffness and does not modify the mass distribution, the analysis is first conducted comparing the response of the two models in the static setting.

The two reference governing equations are clearly given by:

$$\mathbf{K}(a_0)\,\mathbf{v}(a_0) = \mathbf{f} \quad [8.21]$$

for the transverse deflection of the discretized damaged Euler–Bernoulli beam and the fourth-order differential equation of the single-cracked beam expressed by

$$\frac{\tilde{d}^4 v(x,\kappa_0)}{dx^4} - \Delta\Theta_0(\kappa_0)\delta^{(2)}(x-x_0) - P_0\delta(x-\xi) = 0 \quad [8.22]$$

for the continuous model. All the symbols and functions present in equations [8.21] and [8.22] are the same as described for equations [8.11] and [8.14] in dynamic setting. As previously outlined in all the analyses, the considered damaged beams are supposed to be subjected to a concentrated force.

The following analysis takes full advantage of the exact solution of equation [8.22] in terms of deflection $v(x,\kappa_0)$ (as well as all the rotation, bending moment and shear force) provided by Failla (2007, 2011) as explicit function of the spring stiffness κ_0 and not reported for the sake of conciseness. Therefore, the knowledge of the damaged beam deflection via the FE model allows us to evaluate the spring stiffness value $\hat{\kappa}_0$ corresponding to the selected crack depth ratio value $\hat{a}_0$.

Let us consider a single-cracked beam with assigned arbitrary boundary conditions, geometry (namely beam length L and cross-sectional dimensions h and b), material properties (E and ν) and load conditions. The crack position is identified by the proper definition of the stiffness matrix related to the damaged element (see equation [8.4]), and the global stiffness matrix $\mathbf{K}(a_0)$ in equation [8.21] is easily evaluated for a fixed value of the crack depth $a_0 = \hat{a}_0 h$ following the procedure reported in section 8.2.1. Then, the vector $\mathbf{v}(a_0)$ can be straightforwardly calculated.

Due to the explicit expression provided for the deflection function $v(x,\kappa_0)$ for a damaged beam with any boundary condition, by matching the displacements in correspondence of any abscissa of the beam axis (e.g. at the tip $x = L$ for a cantilever beam or at the mid-span beam $x = L/2$ for a simply supported beam) provided by the solutions of equation [8.21] and equation [8.22], respectively, one can immediately find the unknown dimensionless spring stiffness value $\hat{\kappa}_0$. Therefore, for a damaged beam with assigned geometry and material, the comparison of the two models in terms of displacement allows us to evaluate the variation of the dimensionless spring stiffness $\hat{\kappa}_0$ with respect to the dimensionless crack depth $\hat{a}_0$, namely $\hat{\kappa}_0(\hat{a}_0)$.

As an example, Figure 8.2 shows an illustrative curve representing the function $\hat{\kappa}_0(\hat{a}_0)$ obtained for a beam with length L = 8 m, rectangular cross-section with dimensions h = 18 cm and b = 11 cm and Young's modulus $E = 2.1\times10^{11}\,\mathrm{N/m^2}$. Similar curves can be easily evaluated considering different geometry and material for the analyzed damaged beam.

Bearing in mind equations [8.6] and [8.15], the following relationship holds:

$$\kappa_0 = \hat{\kappa}_0(\hat{a}_0)\frac{EI}{L} = \frac{EI}{h}\frac{1}{C_{\hat{\kappa}_0}(\hat{a}_0)} \qquad [8.23]$$

therefore it follows

$$C_{\hat{\kappa}_0}(\hat{a}_0) = \frac{L}{h}\frac{1}{\hat{\kappa}_0(\hat{a}_0)} \qquad [8.24]$$

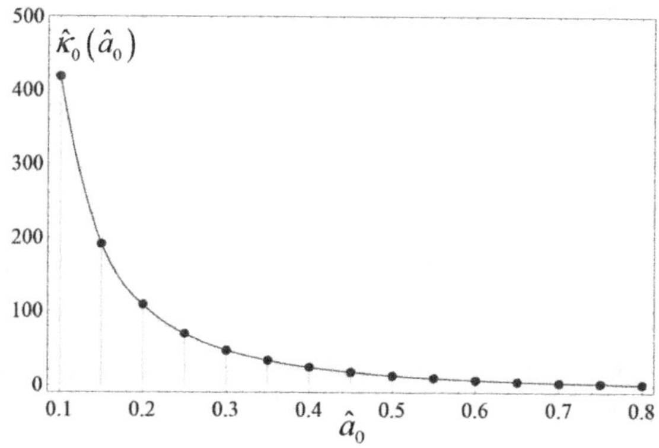

Figure 8.2. *Function $\hat{\kappa}_0(\hat{a}_0)$ versus the dimensionless crack depth $\hat{a}_0$. For a color version of this figure, see www.iste.co.uk/challamel/mechanics3.zip*

Figure 8.3 shows the variation of the compliance function $C_{\hat{\kappa}_0}(\hat{a}_0)$ (see equation [8.24]) versus the dimensionless crack depth $\hat{a}_0$ compared with the curves representing the functions $C(\hat{a}_0)$ previously introduced for the five spring models (see equations [8.16]–[8.20]).

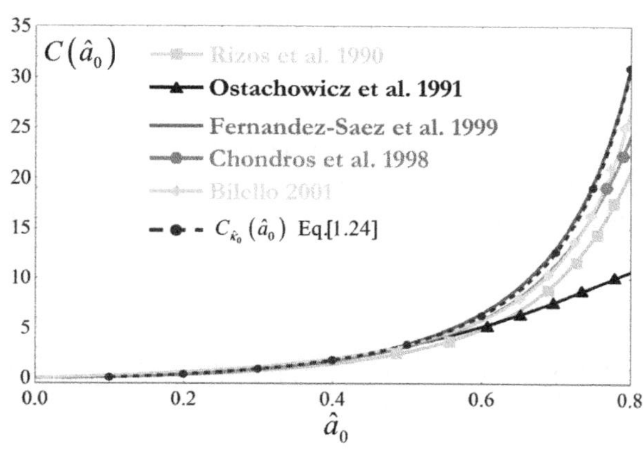

Figure 8.3. *Compliance function $C(\hat{a}_0)$ versus the dimensionless crack depth $\hat{a}_0$. For a color version of this figure, see www.iste.co.uk/challamel/mechanics3.zip*

Note that the proposed compliance function $C_{\hat{\kappa}_0}(\hat{a}_0)$ is in excellent agreement with the function $C(\hat{a}_0)$ provided by Fernandez *et al.* (see equation [8.18]). It is worth noting that for an intact beam ($\hat{a}_0 = a_0 = 0$), no discontinuity is present returning from a physical point of view $\Delta\Theta = 0$. This condition mathematically corresponds to $\kappa_0 \to \infty$, and consequently $C(\hat{a}_0) = 0$. This requirement is satisfied by $C_{\hat{\kappa}_0}(\hat{a}_0)$ together with all the models previously introduced. In the extreme case of $\hat{a}_0 \to 1$ corresponding to the condition of a beam that is completely damaged, physically, the rotational spring ceases to work. The stiffness of the spring $\kappa_0 \to 0$ tends to zero and consequently $C(\hat{a}_0) \to \infty$. This condition is verified for $C_{\hat{\kappa}_0}(\hat{a}_0)$ as well as for the model by Fernandez-Saez *et al.* (equation [8.18]). This requirement is also confirmed in the model proposed by Bilello (see equation [8.20]), while the local compliance functions $C(\hat{a}_0)$ computed by the remaining three models result in finite magnitude quantity. The latter condition implies that the models by Rizos *et al.*, Ostachowicz *et al.* and Chondros *et al.* lose their physical sense for deep cracks.

Given a multi-cracked beam with arbitrary material and boundary conditions, and assigned length L and cross-section height h in the presence of cracks with different depth $a_{0,j} = \hat{a}_{0,j} h$, the corresponding jth dimensionless spring stiffness $\hat{\kappa}_{0,j}$ in the continuous damaged beam model can be easily evaluated by:

$$\hat{\kappa}_{0,j}\left(\hat{a}_{0,j}\right) = \frac{L}{h}\frac{1}{C_{\hat{\kappa}_0}\left(\hat{a}_{0,j}\right)} \qquad [8.25]$$

Let us now consider the case of uncertain crack depth modeled as an uncertain-but-bounded variable as reported in equation [8.9] and rewritten for a single crack (at this step, the subscript j referring to the jth crack is omitted):

$$a^I = a_0(1 + \Delta\alpha\, e_a^I) = \hat{a}_0 h(1 + \Delta\alpha\, e_a^I) \qquad [8.26]$$

As highlighted in section 8.3.1, the corresponding spring stiffness κ^I will be an interval variable too, defined by its lower $\underline{\kappa}$ and upper $\overline{\kappa}$ bounds.

In particular, in terms of dimensionless stiffness $\hat{\kappa}^I = \left[\underline{\hat{\kappa}}, \overline{\hat{\kappa}}\right]$ and bearing in mind equation [8.23], the lower and upper bounds of $\hat{\kappa}(\hat{a}_0)$ can be easily evaluated for selected values of $\hat{a}_0$ and deviation amplitude $\Delta\alpha$ as:

$$\underline{\hat{\kappa}}(\hat{a}_0) = \frac{L}{h}\frac{1}{\overline{C}_{\hat{\kappa}}(\hat{a}_0)} \quad ; \quad \overline{\hat{\kappa}}(\hat{a}_0) = \frac{L}{h}\frac{1}{\underline{C}_{\hat{\kappa}}(\hat{a}_0)} \qquad [8.27]$$

The upper $\overline{C}_{\hat{\kappa}}(\hat{a}_0)$ and lower $\underline{C}_{\hat{\kappa}}(\hat{a}_0)$ bounds of the compliance function in equation [8.27] can be evaluated by following the previous steps. In particular, considering a single-cracked beam and for a selected level of uncertainty $\Delta\alpha$ (see equation [8.26]), it is feasible to build the functions $\underline{\hat{\kappa}}(\hat{a}_0)$ and $\overline{\hat{\kappa}}(\hat{a}_0)$ represented by the curves reported in Figure 8.4(a) and (b) (for $0.2 \leq \hat{a}_0 \leq 0.5$) analogous to that shown in Figure 8.2.

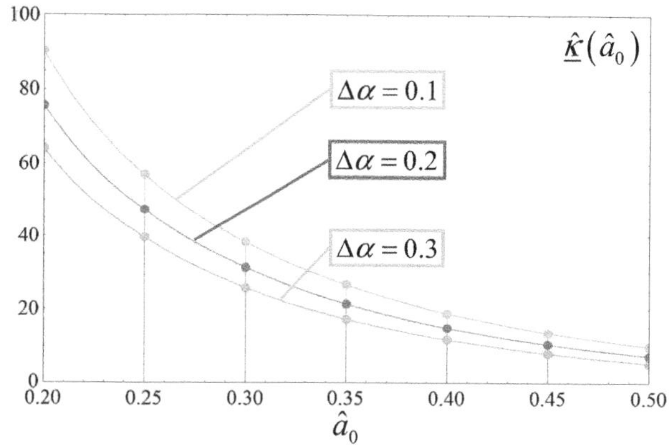

(a)

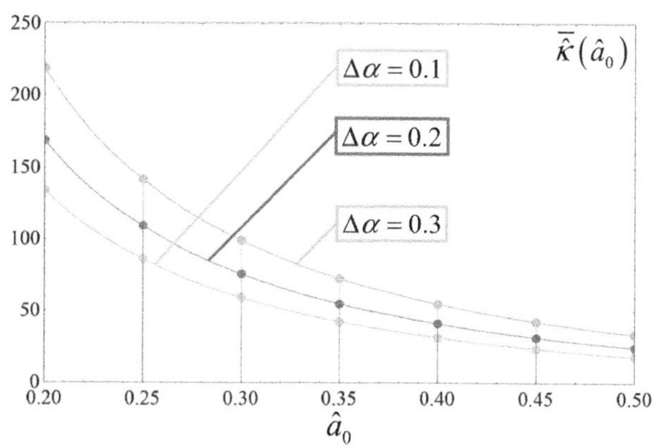

(b)

Figure 8.4. *(a) Lower and (b) upper bounds of the function $\hat{\kappa}(\hat{a}_0)$ for different levels of uncertainty. For a color version of this figure, see www.iste.co.uk/challamel/mechanics3.zip*

Three levels of uncertainty have been considered (namely $\Delta\alpha = 0.1$, $\Delta\alpha = 0.2$ and $\Delta\alpha = 0.3$). Bearing in mind the relationship expressed in equation [8.24], the functions $\underline{C}_{\hat{K}}(\hat{a}_0)$ and $\overline{C}_{\hat{K}}(\hat{a}_0)$ can be straightforwardly evaluated and depicted for the case under examination in Figure 8.5.

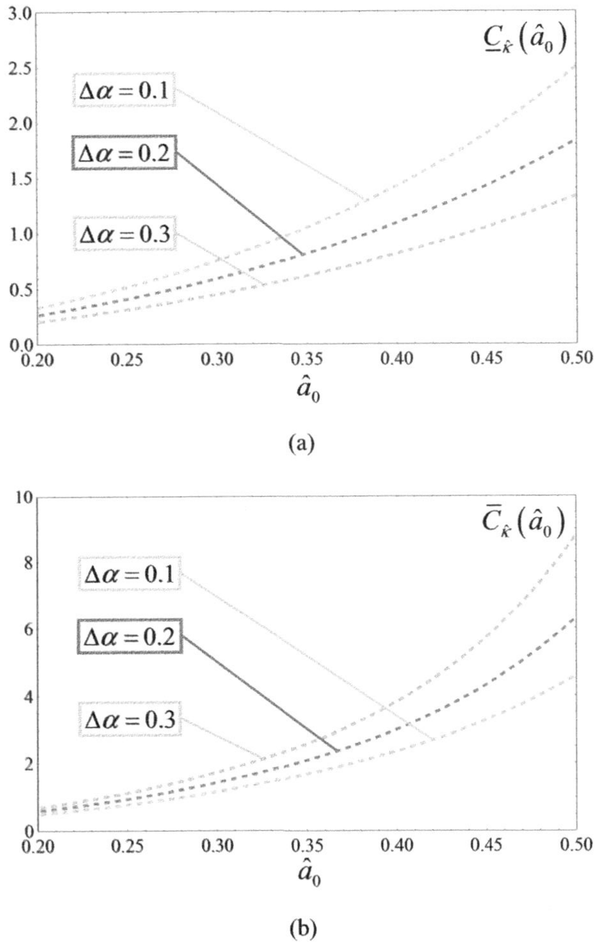

Figure 8.5. *(a) Lower and (b) upper bounds of the function* $C_{\hat{K}}(\hat{a}_0)$ *for different levels of uncertainty. For a color version of this figure, see www.iste.co.uk/challamel/mechanics3.zip*

The functions $\underline{C}_{\hat{\kappa}}(\hat{a}_0)$ and $\overline{C}_{\hat{\kappa}}(\hat{a}_0)$ can be successively used to find the respective bounds $\underline{\hat{\kappa}}_j(\hat{a}_{0,j})$ and $\overline{\hat{\kappa}}_j(\hat{a}_{0,j})$ for a beam with assigned length L and prismatic cross-section with height h in the presence of multiple cracks with dimensionless nominal depth $\hat{a}_{0,j}$ rewriting equation [8.27] in the form:

$$\underline{\hat{\kappa}}_j(\hat{a}_{0,j}) = \frac{L}{h}\frac{1}{\overline{C}_{\hat{\kappa}}(\hat{a}_{0,j})} \quad ; \quad \overline{\hat{\kappa}}_j(\hat{a}_{0,j}) = \frac{L}{h}\frac{1}{\underline{C}_{\hat{\kappa}}(\hat{a}_{0,j})} \qquad [8.28]$$

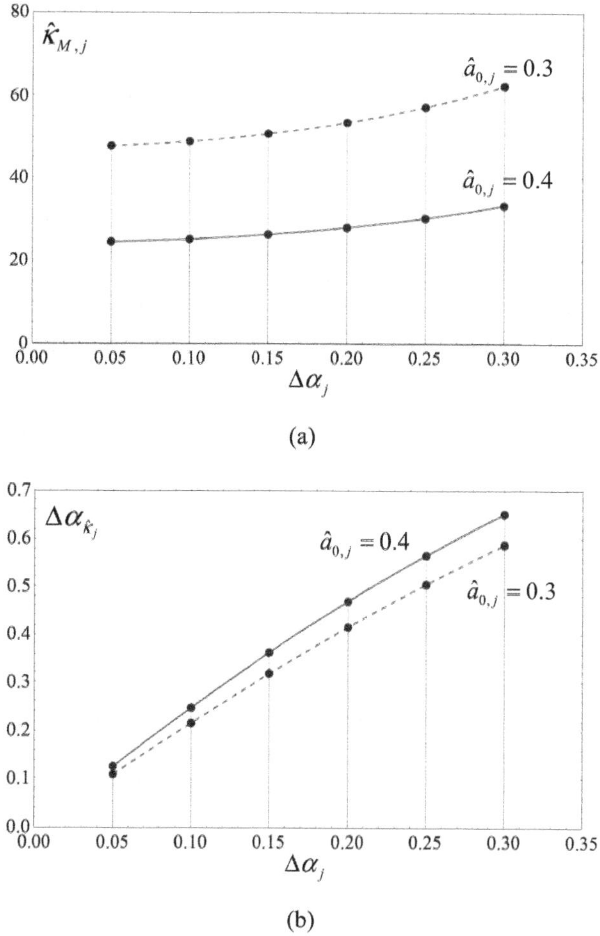

Figure 8.6. *(a) Central value $\hat{\kappa}_{M,j}$ and (b) deviation amplitude $\Delta\alpha_{\hat{\kappa}_j}$ for different levels of uncertainty and two values of nominal crack depth ($a_0 = 0.3h$ and $a_0 = 0.4h$). For a color version of this figure, see www.iste.co.uk/challamel/mechanics3.zip*

The bounds evaluation for the jth interval spring stiffness $\hat{\kappa}_j^I$ allows us to calculate central value $\kappa_{M,j}$ and deviation amplitude $\Delta\hat{\kappa}_j$ (see equations [8.8]) in the form:

$$\hat{\kappa}_{M,j} = \frac{\underline{\hat{\kappa}}_j + \overline{\hat{\kappa}}_j}{2}; \ \Delta\hat{\kappa}_j = \frac{\overline{\hat{\kappa}}_j - \underline{\hat{\kappa}}_j}{2} \quad [8.29]$$

and, following the interval symbolism, in the corresponding affine form:

$$\hat{\kappa}_j^I = \hat{\kappa}_{M,j}(1 + \Delta\alpha_{\hat{\kappa}_j} \hat{e}_{\hat{\kappa},j}^I) \quad [8.30]$$

where $\Delta\alpha_{\hat{\kappa}_j} = \Delta\hat{\kappa}_j / \hat{\kappa}_{M,j}$ and $\hat{\kappa}_{M,j} \neq \kappa_{0,j}$. It is worth noting that in equation [8.30], the central value $\hat{\kappa}_{M,j}$ differs from the nominal value $\kappa_{0,j}$ of the spring stiffness without uncertainty.

In addition, note that equation [8.30] is formally analogous to equation [8.9] for the interval "crack depth" parameter. Moreover, equation [8.30] is directly related to equation [8.9] through equations [8.28] and [8.29], as shown in Figure 8.6, where the central value $\hat{\kappa}_{M,j}$ and the deviation $\Delta\hat{\kappa}_j$ for the jth spring can be evaluated for selected nominal crack depth $\hat{a}_{0,j}$ and level of uncertainty $\Delta\alpha_j$. In particular, Figure 8.6 shows the variation of $\hat{\kappa}_{M,j}$ and $\Delta\alpha_{\hat{\kappa}_j}$ for a level of uncertainty $\Delta\alpha_j$ varying between 0.05 and 0.30 and crack depth nominal values fixed in $\hat{a}_{0,j} = 0.3$ and $\hat{a}_{0,j} = 0.4$.

8.4. Interval frequency response of multi-cracked beams

The next subsections provide interval analytical solutions to the forced vibrations of cracked Euler–Bernoulli beams. Specifically, the bounds of the deflection of the beam steady-state response due to a unit concentrated force acting at an arbitrary position $x = \xi$ (namely the bounds of the dynamic Green's functions) are provided separately for the FE and continuous models.

8.4.1. *Interval deflection function in the FE model*

In this section, the bounds of the deflection function for a discretized multi-cracked beam under a harmonically varying transverse force is evaluated, resorting to a procedure still developed to handle structural problems in the presence of uncertainties modeled as interval variables (Muscolino *et al.* 2014) and formerly applied for damaged beams (Muscolino and Santoro 2017) in the presence of a single crack. The first step is the evaluation in explicit form of the *FRF* in equation [8.13].

The $n \times n$ *interval* stiffness matrix $\mathbf{K}(\mathbf{a}^I)$ in equation [8.13] can be expressed as a function of the uncertain cracks' depths as follows:

$$\mathbf{K}(\mathbf{a}^I) = \mathbf{K}(\mathbf{a}_0, \boldsymbol{\alpha}) = \mathbf{K}_C(\mathbf{a}_0) + \sum_{j=1}^{N_c} \mathbf{K}_j(\mathbf{a}_0) \Delta\alpha_j \, e_{a,j}^I \qquad [8.31]$$

with

$$\mathbf{K}_C(\mathbf{a}_0) = \mathbf{K}(\boldsymbol{\alpha}^I)\Big|_{\boldsymbol{\alpha}=\mathbf{0}} \quad ; \quad \mathbf{K}_j(\mathbf{a}_0) = \left.\frac{\partial \mathbf{K}(\boldsymbol{\alpha}^I)}{\partial \alpha_j}\right|_{\boldsymbol{\alpha}=\mathbf{0}} \qquad [8.32]$$

In equation [8.32], $\mathbf{K}_C(\mathbf{a}_0)$ is a positive definite symmetric matrix of order $n \times n$, representing the *nominal stiffness matrix* referred to the damaged beam without uncertainty in the cracks' depths (see equation [8.10]), while $\mathbf{K}_j(\mathbf{a}_0)$ is a symmetric matrix of order $n \times n$ and rank r. Specifically, the latter can be expressed as a superposition of rank-one matrices such as

$$\mathbf{K}_j(\mathbf{a}_0) = \sum_{k=1}^{r} \lambda_j^{(k)} \mathbf{w}_j^{(k)} \mathbf{w}_j^{(k)T} \qquad [8.33]$$

with $\mathbf{w}_j^{(k)} = \mathbf{K}_C(\mathbf{a}_0)\boldsymbol{\psi}_j^{(k)}$, where $\boldsymbol{\psi}_j^{(k)}$ and $\lambda_j^{(k)}$ are, respectively, the kth eigenvector $(k = 1, \ldots, r)$ and the associated eigenvalue solutions of the following eigenproblem:

$$\begin{aligned} &\mathbf{K}_j(\mathbf{a}_0)\boldsymbol{\psi}_j^{(k)} = \lambda_j^{(k)} \mathbf{K}_C(\mathbf{a}_0)\boldsymbol{\psi}_j^{(k)}; \quad \boldsymbol{\Psi}_j^T \mathbf{K}_C(\mathbf{a}_0)\boldsymbol{\Psi}_j = \mathbf{I}_r; \\ &\boldsymbol{\Psi}_j = \left[\boldsymbol{\psi}_j^{(1)} \quad \boldsymbol{\psi}_j^{(2)} \quad \cdots \quad \boldsymbol{\psi}_j^{(r)}\right], \quad (j = 1, \ldots, N_c; k = 1, \ldots, r) \end{aligned} \qquad [8.34]$$

Note that only $r < n$ eigenvalues are different from zero.

Therefore, taking into account the decomposition of the stiffness matrix (see equation [8.31]), equation [8.12] can be rewritten as

$$\mathbf{V}(\mathbf{a}^I, \omega) = \left[\mathbf{I}_n + \mathbf{H}_C(\mathbf{a}_0, \omega)\, \mathbf{S}^I(\boldsymbol{\alpha}, \omega)\right]^{-1} \mathbf{H}_C(\mathbf{a}_0, \omega)\mathbf{F}(\omega) \qquad [8.35]$$

In the previous equation, $\mathbf{I}_n$ denotes the identity matrix of order *n*, $\mathbf{H}_\mathrm{C}(\mathbf{a}_0,\omega)$ is the *FRF* matrix of the nominal damaged beam referred to as the *nominal stiffness matrix* and $\mathbf{S}^I(\boldsymbol{\alpha})$ is a $n\times n$ matrix accounting for the fluctuations of the uncertain parameters. $\mathbf{H}_\mathrm{C}(\mathbf{a}_0,\omega)$ and $\mathbf{S}^I(\boldsymbol{\alpha})$ are given, respectively, by:

$$\mathbf{H}_\mathrm{C}(\mathbf{a}_0,\omega)=\left[-\omega^2\,\mathbf{M}+\mathbf{K}_\mathrm{C}(\mathbf{a}_0)\right]^{-1}\;;\;\mathbf{S}^I(\boldsymbol{\alpha})=\sum_{j=1}^{N_c}\sum_{k=1}^{r}\lambda_j^{(k)}\mathbf{w}_j^{(k)}\mathbf{w}_j^{(k)T}\Delta\alpha_j\,e_{a,j}^I \tag{8.36}$$

The nominal *FRF* $\mathbf{H}_\mathrm{C}(\mathbf{a}_0,\omega)$ can be evaluated in closed form as:

$$\mathbf{H}_\mathrm{C}(\mathbf{a}_0,\omega)=\boldsymbol{\Phi}_\mathrm{C}\,\mathbf{H}_{\mathrm{C,m}}(\mathbf{a}_0,\omega)\boldsymbol{\Phi}_\mathrm{C}^T=\boldsymbol{\Phi}_\mathrm{C}\left[-\omega^2\mathbf{I}_m+\boldsymbol{\Omega}_\mathrm{C}^2\right]^{-1}\boldsymbol{\Phi}_\mathrm{C}^T \tag{8.37}$$

where $\boldsymbol{\Phi}_\mathrm{C}$ is the modal matrix, of order $n\times m$, pertaining to the *mean* configuration evaluated as the solution of the following eigenproblem:

$$\mathbf{K}_\mathrm{C}\boldsymbol{\Phi}_\mathrm{C}=\mathbf{M}\,\boldsymbol{\Phi}_\mathrm{C}\,\boldsymbol{\Omega}_\mathrm{C}^2;\quad\boldsymbol{\Phi}_\mathrm{C}^T\,\mathbf{M}\boldsymbol{\Phi}_\mathrm{C}=\mathbf{I}_m;\quad\boldsymbol{\Phi}_\mathrm{C}^T\,\mathbf{K}_\mathrm{C}\,\boldsymbol{\Phi}_\mathrm{C}=\boldsymbol{\Omega}_\mathrm{C}^2 \tag{8.38}$$

with $\boldsymbol{\Omega}_\mathrm{C}^2$ being the spectral matrix listing the squares of the natural circular frequencies of the nominal damaged beam and $\mathbf{I}_m$ the identity matrix of order *m*.

By inspecting equations [8.35] and [8.36], it is observed that the evaluation of the *FRF* $\mathbf{H}(\mathbf{a}^I,\omega)$ matrix involves the inversion of a matrix expressed as a sum of a *mean nominal FRF* matrix plus a deviation given as superposition of rank-one matrices. Therefore, the *FRF* can be evaluated in approximate explicit form by applying the *interval rational series expansion* (*IRSE*) (Muscolino *et al.* 2014), herein truncated to first-order terms:

$$\mathbf{H}^I(\omega)\approx\mathbf{H}_\mathrm{C}(\mathbf{a}_0,\omega)-\sum_{j=1}^{N_c}\sum_{k=1}^{r}\frac{\lambda_j^{(k)}\Delta\alpha_j\,\hat{e}_{\alpha_j}^I}{1+\Delta\alpha_j\,\hat{e}_{\alpha_j}^I\,\lambda_j^{(k)}\,g_{jk}(\omega)}\mathbf{G}_{jk}(\omega)=\mathbf{H}_\mathrm{mid}(\omega)+\mathbf{H}_\mathrm{dev}^I(\omega) \tag{8.39}$$

with

$$g_{jk}(\omega)=\mathbf{w}_j^{(k)T}\mathbf{H}_\mathrm{C}(\mathbf{a}_0,\omega)\;\mathbf{w}_j^{(k)};\quad\mathbf{G}_{jk}(\omega)=\mathbf{H}_\mathrm{C}(\mathbf{a}_0,\omega)\,\mathbf{w}_j^{(k)}\mathbf{w}_j^{(k)T}\mathbf{H}_\mathrm{C}(\mathbf{a}_0,\omega) \tag{8.40}$$

Equation [8.39] provides the *interval FRF* matrix as a sum of the midpoint matrix, $\mathbf{H}_\mathrm{mid}(\omega)$, plus the interval deviation matrix, $\mathbf{H}_\mathrm{dev}^I(\omega)$, given, respectively, by:

$$\mathbf{H}_{\text{mid}}(\omega) = \mathbf{H}_{\text{C}}(\mathbf{a}_0, \omega) + \sum_{j=1}^{N_c}\sum_{k=1}^{r} q_{0,jk}(\omega)\mathbf{G}_{jk}(\omega) \quad [8.41]$$

and

$$\mathbf{H}_{\text{dev}}^{I}(\omega) = \sum_{j=1}^{N_c}\sum_{k=1}^{r} \Delta q_{jk}(\omega)\mathbf{G}_{jk}(\omega)\hat{e}_{\alpha_j}^{I} \quad [8.42]$$

with

$$q_{0,jk}(\omega) = \frac{(\lambda_j^{(k)}\Delta\alpha_j)^2 g_{jk}(\omega)}{1-\left[\lambda_j^{(k)}\Delta\alpha_j\, g_{jk}(\omega)\right]^2}; \quad \Delta q_{jk}(\omega) = \frac{\lambda_j^{(k)}\Delta\alpha_j}{1-\left[\lambda_j^{(k)}\Delta\alpha_j\, g_{jk}(\omega)\right]^2} \quad [8.43]$$

Then, the lower and upper bounds of the frequency response can be, respectively, evaluated as

$$\begin{aligned}\underline{\mathbf{V}}(\omega) &= \min\left\{\mathbf{V}_{mid}(\omega) - \left|\mathbf{V}_{dev}(\omega)\right|, \mathbf{V}_{mid}(\omega) + \left|\mathbf{V}_{dev}(\omega)\right|\right\};\\ \overline{\mathbf{V}}(\omega) &= \max\left\{\mathbf{V}_{mid}(\omega) - \left|\mathbf{V}_{dev}(\omega)\right|, \mathbf{V}_{mid}(\omega) + \left|\mathbf{V}_{dev}(\omega)\right|\right\}\end{aligned} \quad [8.44]$$

where in equation [8.44], the following positions hold:

$$\mathbf{V}_{\text{mid}}(\omega) = \mathbf{H}_{\text{mid}}(\omega)\mathbf{F}(\omega) \;\; ; \; \mathbf{V}_{dev}(\omega) = \left(\sum_{j=1}^{N_c}\sum_{k=1}^{r} \Delta q_{jk}(\omega)\mathbf{G}_{jk}(\omega)\right)\mathbf{F}(\omega) \quad [8.45]$$

8.4.2. *Interval deflection function in the continuous model*

Following the approach presented in Failla (2009, 2011), the frequency deflection response $V\left(x,\omega,\boldsymbol{\kappa}^I\right)$ governed by equation [8.14] can be obtained in exact analytical form with an explicit dependence by the uncertainties collected in the vector $\boldsymbol{\kappa}^I = [\kappa_1^J \;\kappa_2^J \ldots \kappa_{N_c}^J]^T$. As previously outlined, the definition of the stiffness of the springs as uncertain-but-bounded variables implies the frequency response deflection function turns out to be an interval function too.

Lower and upper bounds of the frequency deflection function for a multi-cracked continuous beam under a harmonically varying transverse force are calculated via a method recently developed by an author specializing in static settings and whose knowledge extends to dynamics (Santoro *et al.* 2020).

Specifically if, for a given frequency ω, the deflection function $V\left(x,\omega,\boldsymbol{\kappa}^I\right)$ is monotonic with respect to all the uncertain parameters (see Santoro *et al.* 2020), the lower and upper bounds of $V\left(x,\omega,\boldsymbol{\kappa}^I\right)$ can be evaluated performing a typical sensitivity-based approach.

The calculation of the sensitivity functions $s_{\kappa_j}^{(V)}(x)$ setting all the uncertain parameters to the respective nominal values as

$$s_{\kappa_j}^{(V)}(x) = \left.\frac{\partial V(x,\omega,\boldsymbol{\kappa})}{\partial \kappa_j}\right|_{\boldsymbol{\kappa}=\boldsymbol{\kappa}_0} \qquad \underline{\kappa}_j \le \kappa_j \le \overline{\kappa}_j \tag{8.46}$$

allows us to fix a correspondence between the bounds of every uncertain parameter and the bounds of the response for that parameter. In detail, if the sign of the sensitivity function in equation [8.46] is positive, the lower bounds/upper bounds of the parameter will determine lower bounds/upper bounds of the response; the correspondence is inverted if the sign is negative, i.e. the lower bounds/upper bounds of the parameter will determine the upper bounds/lower bounds of the response.

Essentially, the extreme values of the uncertain parameters are evaluated by:

$$\kappa_j^{(LB)} = \underline{\kappa}_j, \quad \kappa_j^{(UB)} = \overline{\kappa}_j \text{ for } s_{\kappa_j}^{(V)}(x) > 0 \tag{8.47}$$

$$\kappa_j^{(LB)} = \overline{\kappa}_j, \quad \kappa_j^{(UB)} = \underline{\kappa}_j \text{ for } s_{\kappa_j}^{(V)}(x) < 0 \tag{8.48}$$

and the bounds of the deflection function are calculated as:

$$\underline{V}(x,\omega) = V\left(x,\omega,\boldsymbol{\kappa}^{(LB)}\right); \quad \overline{V}(x,\omega) = V\left(x,\omega,\boldsymbol{\kappa}^{(UB)}\right) \tag{8.49}$$

where the vectors

$$\boldsymbol{\kappa}^{(LB)} = \begin{bmatrix} \kappa_1^{(LB)} & \kappa_1^{(LB)} & \cdots & \kappa_{N_c}^{(LB)} \end{bmatrix}^T ; \boldsymbol{\kappa}^{(UB)} = \begin{bmatrix} \kappa_1^{(UB)} & \kappa_1^{(UB)} & \cdots & \kappa_{N_c}^{(UB)} \end{bmatrix}^T \tag{8.50}$$

collect the components obtained by equations [8.47] or [8.48].

8.5. Numerical applications

The analysis is performed on a damaged steel simply supported beam subjected to a time harmonic concentrated force $P_0 \exp(i\omega t)$ applied at $x=\xi$ (see Figure 8.1). The beam has length L = 8 m and a rectangular cross-section with width b = 0.11 m and height h = 0.18 m. The Young modulus and the material mass density are selected as $E = 2.1\times10^{11}\,\mathrm{N/m^2}$ and $\rho = 7850\,\mathrm{Kg/m^3}$, respectively.

Let us assume positive upward vertical displacements and counter-clockwise rotations.

The results are reported in terms of lower and upper bounds of the deflection function for 18 Hz unitary point load (P_0 = 1N) applied at ξ = 4.8 m. Two damage scenarios are considered. The first damage configuration involves $N_c = 3$ cracks at positions $x_1 = 2$ m, $x_2 = 3.6$ m and $x_3 = 4.4$ m. The second one takes into account $N_c = 6$ cracks with three additional cracks located at $x_4 = 6$ m, $x_5 = 6.8$ m and $x_6 = 7.6$ m.

Let us first take into consideration the damaged beam with $N_c = 3$ cracks. Based on the FE approach, the simply supported beam has been modeled by $N_e = 30$ finite elements with the cracks supposed to be located in the middle of the 8th, 14th and 17th elements. The nominal values of the cracks' depths are assumed to be $a_{0,1} = 0.3h$ and $a_{0,2} = a_{0,3} = 0.4h$ ($\hat{a}_{0,1} = 0.3$, $\hat{a}_{0,2} = \hat{a}_{0,3} = 0.4$). The deviations from the nominal values are assumed to be $\Delta\alpha_1 = \Delta\alpha_3 = 0.3$ and $\Delta\alpha_2 = 0.2$ (see equation [8.9]).

Following the procedure described in section 8.4.1, the bounds of the deflection function are evaluated via equation [8.44]. Figure 8.7 shows the lower and upper bounds calculated for the case under examination, and the accuracy of the proposed method is confirmed by comparing with the displacement bounds built by the combinatorial vertex method (Dong and Shah 1987), which represents the reference exact solution. It is worth noting that the vertex exact solution requires the evaluation of $2^{N_c=3} = 8$ analyses at every abscissa x to explore all possible combinations arising between the lower and upper bounds of the uncertain parameters.

In order to evaluate the deflection function bounds resorting to the continuous damaged beam model, first let us consider the correspondence between the crack depth and the spring stiffness, extensively described in section 8.3.3.

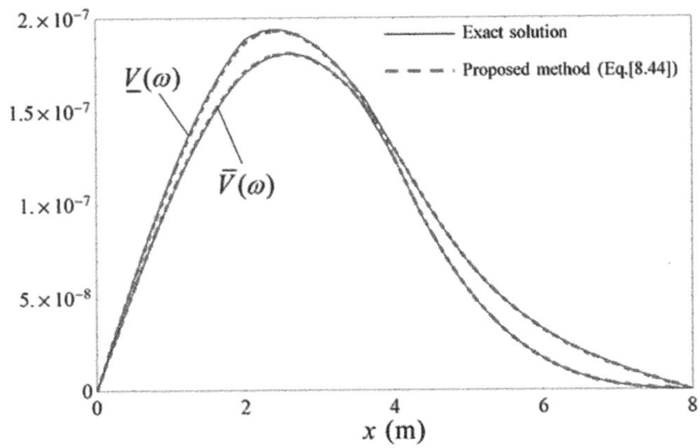

Figure 8.7. *Lower and upper bounds of the deflection function (in m) for a finite element multi-cracked (N_c = 3) simply supported beam with uncertain-but-bounded crack depths subjected to a unitary point load (ω= 18 Hz). For a color version of this figure, see www.iste.co.uk/challamel/mechanics3.zip*

In particular, taking full advantage of the relationships in equations [8.28]–[8.30] as well as of the curves reported in Figure 8.6, the central values and the deviations for the stiffness of the $N_c = 3$ springs can be easily obtained. For the case under examination, central and deviation values of the first spring are, respectively, $\hat{\kappa}_{M,1} = 62.42$ and $\Delta\alpha_{\hat{\kappa}_1} = 0.59$ (corresponding to $a_{0,1} = 0.3h$ and $\Delta\alpha_1 = 0.3$); the values calculated for the second spring are $\hat{\kappa}_{M,2} = 27.93$ and $\Delta\alpha_{\hat{\kappa}_2} = 0.47$ (corresponding to $a_{0,2} = 0.4h$ and $\Delta\alpha_2 = 0.2$) and finally $\hat{\kappa}_{M,3} = 33.24$ and $\Delta\alpha_{\hat{\kappa}_3} = 0.65$ (corresponding to $a_{0,3} = 0.4h$ and $\Delta\alpha_3 = 0.3$) for the third spring.

The interval expressions for all the uncertain parameters as reported in equation [8.30] are well defined for the continuous damaged beam. Once the stiffness of the $N_c = 3$ springs is calculated (see equation [8.6]), the deflection function bounds can be evaluated performing the procedure reported in section 8.4.2.

In Figure 8.8(a), the sensitivities $s_{\kappa_j}^{(V)}(x)$ for $j = 1,...,3$ as defined in equation [8.46] are shown versus the beam axis x, and their trends (in sign) allow us to evaluate the components of vectors $\boldsymbol{\kappa}^{(LB)}$ and $\boldsymbol{\kappa}^{(UB)}$ (see equation [8.50]) at every abscissa x. In Figure 8.8(b), the deflection bounds built by applying equation [8.49] are shown and compared again with the reference solution provided by the vertex method. Note again the excellent agreement between the proposed and exact solutions.

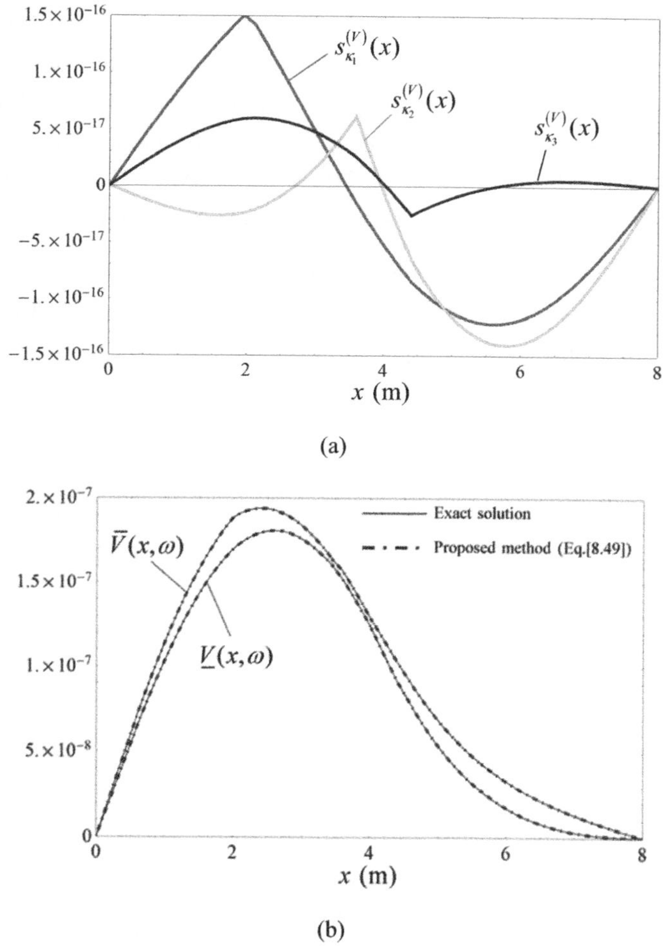

Figure 8.8. *(a) Sensitivity functions and (b) lower and upper bounds of the deflection function (in m) for a continuous multi-cracked (N_c = 3) simply supported beam with uncertain-but-bounded stiffness of springs subjected to a unitary point load (ω= 18 Hz). For a color version of this figure, see www.iste.co.uk/challamel/mechanics3.zip*

The same procedures are adopted to evaluate the interval deflection response for the second damage configuration involving $N_c = 6$ cracks. For the FE model, a uniform mesh of $N_e = 30$ finite elements is again considered with the cracks located in the middle of the 8th, 14th, 17th, 23rd, 26th and 29th elements. The nominal values of the cracks' depths are assumed to be $a_{0,1} = a_{0,5} = a_{0,6} = 0.3h$ and $a_{0,2} = a_{0,3} = a_{0,4} = 0.4h$ ($\hat{a}_{0,1} = \hat{a}_{0,5} = \hat{a}_{0,6} = 0.3$, $\hat{a}_{0,2} = \hat{a}_{0,3} = \hat{a}_{0,4} = 0.4$), while the deviations from the nominal values are fixed in $\Delta\alpha_1 = \Delta\alpha_3 = \Delta\alpha_5 = \Delta\alpha_6 = 0.3$ and $\Delta\alpha_2 = \Delta\alpha_4 = 0.2$ (see equation [8.9]). The corresponding central and deviation values for the $N_c = 6$ springs are straightforwardly calculated. Specifically for the first, fifth and sixth springs central values and deviations are evaluated in $\hat{\kappa}_{M,1} = \hat{\kappa}_{M,5} = \hat{\kappa}_{M,6} = 62.42$ and $\Delta\alpha_{\hat{\kappa}_1} = \Delta\alpha_{\hat{\kappa}_5} = \Delta\alpha_{\hat{\kappa}_6} = 0.59$, respectively (corresponding to $a_{0,1} = a_{0,5} = a_{0,6} = 0.3h$ and $\Delta\alpha_1 = \Delta\alpha_5 = \Delta\alpha_6 = 0.3$); the second and fourth springs are characterized by $\hat{\kappa}_{M,2} = \hat{\kappa}_{M,4} = 27.93$ and $\Delta\alpha_{\hat{\kappa}_2} = \Delta\alpha_{\hat{\kappa}_4} = 0.47$ (corresponding to $a_{0,2} = a_{0,4} = 0.4h$ and $\Delta\alpha_2 = \Delta\alpha_4 = 0.2$) and the third spring is characterized again by the values $\hat{\kappa}_{M,3} = 33.24$ and $\Delta\alpha_{\hat{\kappa}_3} = 0.65$ (corresponding to $a_{0,3} = 0.4h$ and $\Delta\alpha_3 = 0.3$).

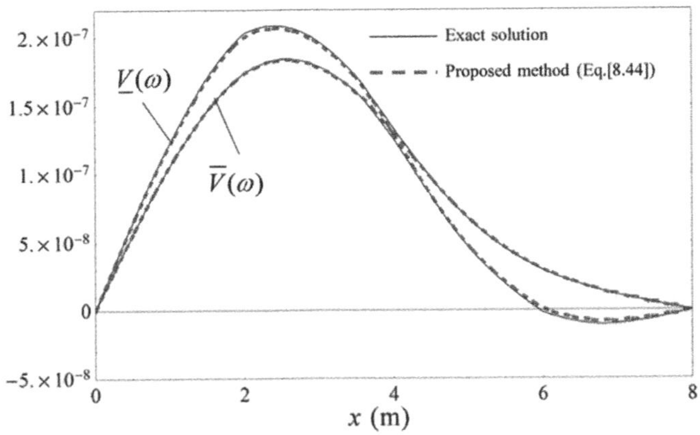

Figure 8.9. *Lower and upper bounds of the deflection function (in* m*) for a finite element multi-cracked (N_c = 6) simply supported beam with uncertain-but-bounded crack depths subjected to a unitary point load (ω = 18 Hz). For a color version of this figure, see www.iste.co.uk/challamel/mechanics3.zip*

Results in terms of deflection function bounds for the FE model and for the continuous model are reported in Figures 8.9 and 8.10, respectively. In particular, Figure 8.10(a) shows the trends of the sensitivity functions essential for the evaluations of the bounds as expressed in equation [8.49].

For both damage models, the proposed interval solutions are compared with the reference exact solution involving, for the second damage scenario, $2^{N_c=6} = 64$ analyses at every abscissa x. Despite the number of cracks considered and the high level of uncertainty, the proposed solutions show again an excellent agreement with the exact solutions, especially by the procedure proposed for the continuous model.

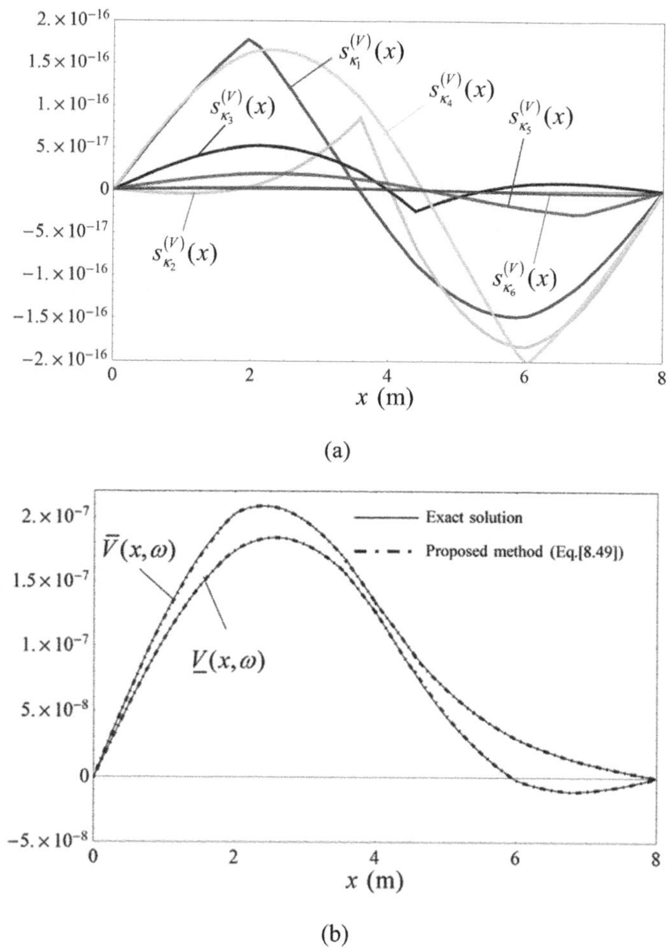

Figure 8.10. *(a) Sensitivity functions and (b) lower and upper bounds of the deflection function for a continuous multi-cracked (N_c = 6) simply supported beam with uncertain-but-bounded stiffness of springs subjected to a unitary point load (ω= 18 Hz). For a color version of this figure, see www.iste.co.uk/challamel/mechanics3.zip*

To further highlight the accuracy of the proposed methods, an error measure is applied. Specifically, the "*error in bound*" $\varepsilon_b\%$ estimated as

$$\varepsilon_b\% = \left(\frac{|Computed\ bound - Exact\ bound|}{Exact\ bound} \right) \times 100 \qquad [8.51]$$

which provides the percentage error in the lower and upper bounds evaluated via the proposed procedures compared with those calculated via the combinatorial exact solution.

Specifically, for the cases under examination, the "error in bound" $\varepsilon_b\%$ is evaluated in correspondence with the abscissa $x = 2.4$ m where a larger amplitude of the interval response is observed.

The results reported in Table 8.1 show that the procedure evaluated to deal with cracked beams by continuous model provides almost exact deflection bounds values regardless of the order of uncertainty and the number of cracks considered without requiring the assessment of 2^{Nc} evaluations of the combinatorial method. The interval procedure presented for the analysis of multi-cracked beams loses in accuracy by increasing the number of cracks; however, it still provides solutions with reasonable sharpness for these levels of uncertainty.

	$N_c = 3$		$N_c = 6$	
	Lower	**Upper**	**Lower**	**Upper**
Exact	1.7977×10^{-7}	1.9393×10^{-7}	1.833×10^{-7}	2.0861×10^{-7}
Proposed equation [8.44]	1.8012×10^{-7}	1.9333×10^{-7}	1.833×10^{-7}	2.0861×10^{-7}
$\varepsilon_b\%$	**0.19%**	**0.32%**	**0.4%**	**0.9%**
Proposed equation [8.49]	1.7977×10^{-7}	1.9393×10^{-7}	1.834×10^{-7}	2.0861×10^{-7}
$\varepsilon_b\%$	**0%**	**0%**	**0.04%**	**0%**

Table 8.1. *Deflection bounds (in m) and "error in bound" $\varepsilon_b\%$ evaluated at x = 2.4 m for a "finite element" and "continuous" multi-cracked simply supported beam with uncertain-but-bounded crack depths (Nc = 3 and Nc = 6) subjected to a unitary point load (ω= 18 Hz)*

Two remarks arise at this step. The interval solution provided by the method developed for the continuous damaged beam model is not affected by the approximation due to the truncation for the *FRF* matrix, as reported in equation [8.39] in section 8.4.1. However, thanks to the graphics in Figures 8.7, 8.8, 8.9 and 8.10, it is evident that both the proposed procedures allow us to build with great accuracy the interval frequency response for damaged beams with uncertain-but-bounded uncertainties in "*finite element*" and "*continuous*" approaches without suffering from a computational point of view due to the number of uncertain parameters considered, as in the case of the combinatorial vertex method.

8.6. Concluding remarks

This chapter has dealt with the frequency response of the Euler–Bernoulli beams with multiple open cracks affected by uncertainty. The damage has been represented by the local flexibility model distinguishing the finite element approach and the continuous one. Following the two alternative approaches, the generic crack has been modeled building an *ad hoc* cracked finite element with reduced stiffness or by a linearly elastic rotational spring. In the first case, the uncertain parameter is the crack depth while in the second case is the stiffness of the spring.

Based on a non-probabilistic approach, the uncertainty has been represented resorting to the interval model. A correspondence between the crack depth and the spring stiffness has been presented in terms of respective bounds evaluating the interval compliance function.

Regarding the two models, two different procedures have been presented to handle multi-cracked beams with uncertain-but-bounded parameters.

Results obtained in terms of deflection function bounds for a simply supported damaged beam show the efficiency and accuracy of both the present methods as well as the relationship relating the interval crack depth with the interval spring stiffness.

8.7. Acknowledgments

The author wishes to express her deep gratitude to Professor Isaac Elishakoff for his valuable mentorship which encouraged her professional and personal development.

8.8. References

Bilello, C. (2001). Theoretical and experimental investigation on damaged beams under moving systems. PhD Thesis. Università degli Studi di Palermo, Italy.

Cacciola, P. and Muscolino, G. (2002). Dynamic response of a rectangular beam with a known non-propagating crack of certain and uncertain depth. *Comput. Struct.*, 80, 2387–2396.

Caddemi, S. and Caliò, I. (2009). Exact closed-form solution for the vibration modes of the Euler–Bernoulli beam with multiple open cracks. *J. Sound Vib.*, 327(3–5), 473–489.

Caddemi, S. and Morassi, A. (2013). Multi-cracked Euler–Bernoulli beams: Mathematical modeling and exact solutions. *Int. J. Solids Struct.*, 50(6), 944–956.

Chaudhari, T.D. and Maiti, S.K. (2000). A study of vibration of geometrically segmented beams with and without crack. *Int. J. Solids Struct.*, 37(5), 761–779.

Chondros, T., Dimarogonas, A., Yao, J. (1998). A continuous cracked beam vibration theory. *J. Sound Vib.*, 215, 17–34.

Dong, W. and Shah, H.C. (1987). Vertex method for computing functions of fuzzy variables. *Fuzzy Set Syst.*, 24(1), 65–78.

Elishakoff, I. (1995). Essay on uncertainties in elastic and viscoelastic structures: From A.M. Freudenthal's criticisms to modern convex modelling. *Comput. Struct.*, 17(6), 871–895.

Elishakoff, I. and Miglis, Y. (2012). Overestimation-free computational version of interval analysis. *Int. J. Comput. Methods Eng. Sci. Mech.*, 13(5), 319–328.

Elishakoff, I. and Ohsaki, M. (2010). *Optimization and Anti-Optimization of Structures Under Uncertainty*. Imperial College Press, London.

Failla, G. (2011). Closed-form solutions for Euler–Bernoulli arbitrary discontinuous beams. *Arch. Appl. Mech.*, 81(5), 605–628.

Failla, G. (2016). An exact generalised function approach to frequency response analysis of beams and plane frames with the inclusion of viscoelastic damping. *J. Sound Vib.*, 360, 171–202.

Failla, G. (2019). An exact modal analysis approach to vibration analysis of structures with mass-spring subsystems and rotational joints. *J. Sound Vib.*, 438, 191–219.

Failla, G. and Santini, A. (2007). On Euler–Bernoulli discontinuous beam solutions via uniform-beam Green's functions. *Int. J. Solids Struct.*, 44(22–23), 7666–7687.

Falsone, G. (2002). The use of generalised functions in the discontinuous beam bending differential equation. *Int. J. Eng. Educ.*, 18(3), 337–343.

Fernandez-Saez, J., Rubio, L., Navarro, C. (1999). Approximate calculation of the fundamental frequency for bending vibrations of cracked beams. *J. Sound Vib.*, 225, 345–352.

Moens, D. and Vandepitte, D. (2005). A survey of non-probabilistic uncertainty treatment in finite element analysis. *Comput. Methods Appl. Mech. Eng.*, 194, 1527–1555.

Moore, R.E. (1966). *Interval Analysis*. Prentice-Hall, Englewood Cliffs.

Muhanna, R.L. and Mullen, R.L. (2001). Uncertainty in mechanics: Problems – interval – based approach. *J. Eng. Mech. ASCE*, 127, 557–66.

Muscolino, G. and Santoro, R. (2017a). Dynamic response of damaged beams with uncertain crack depth. *AIMETA 2017 – Proceedings of the 23rd Conference of the Italian Association of Theoretical and Applied Mechanics*, 3, 2385–2397.

Muscolino, G. and Santoro, R. (2017b). Explicit frequency response function of beams with crack of uncertain depth. *Procedia Eng.*, 199, 1128–1133.

Muscolino, G. and Santoro, R. (2019). Dynamics of multiple cracked prismatic beams with uncertain-but-bounded depths under deterministic and stochastic loads. *J. Sound Vib.*, 443, 717–731.

Muscolino, G. and Sofi, A. (2012). Stochastic response of structures with uncertain-but-bounded parameters via improved interval analysis. *Probab. Eng. Mech.*, 28, 152–163.

Muscolino, G., Santoro, R., Sofi, A. (2014). Explicit frequency response functions of discretized structures with uncertain parameters. *Comput. Struct.*, 133, 64–78.

Ostachowicz, W. and Krawczuk, M. (1991). Analysis of the effect of cracks on the natural frequencies of a cantilever beam. *J. Sound Vib.*, 150, 191–201.

Qian, G.L., Gu, S.N., Jiang, J.S. (1990). The dynamic behaviour and crack detection of a beam with a crack. *J. Sound Vib.*, 138(2), 233–43.

Qiu, Z.P. and Wang, X.J. (2003). Comparison of dynamic response of structures with uncertain-but-bounded parameters using non probabilistic interval analysis method and probabilistic approach. *Int. J. Solids Struct.*, 40(20), 5423–5439.

Qiu, Z. and Wang, X.J. (2005). Parameter perturbation method for dynamic response of structures with uncertain-but-bounded parameters based on interval analysis. *Int. J. Solids Struct.*, 42(18–19), 4958–4970.

Rizos, P., Aspragathos, N., Dimarogonas, A. (1990). Identification of crack location and magnitude in a cantilever beam from the vibration modes. *J. Sound Vib.*, 138, 381–388.

Ruotolo, R., Surace, C., Crespo, P., Storer, D. (1996). Harmonic analysis of the vibrations of a cantilevered beam with a closing crack. *Comput. Struct.*, 6, 1057–1074.

Santoro, R. and Failla, G. (2021). An interval framework for uncertain frequency response of multi-cracked beams with application to vibration reduction via tuned mass dampers, *Meccanica* (in press). DOI: 10.1007/s11012-020-01290-3.

Santoro, R. and Muscolino, G. (2019). Dynamics of beams with uncertain crack depth: Stochastic versus interval analysis. *Meccanica*, 54(9), 1433–1449.

Santoro, R., Failla, G., Muscolino, G. (2020). Interval static analysis of multi-cracked beams with uncertain size and position of cracks. *Appl. Math. Model.*, 86, 92–114.

Yavari, A. and Sarkani, S. (2001). On applications of generalized functions to the analysis of Euler–Bernoulli beam–columns with jump discontinuities. *Int. J. Mech. Sci.*, 43, 1543–1562.

Zhao, X., Zhao, Y.R., Gao, X.Z., Li, X.Y., Li, Y.H. (2016). Green's functions for the forced vibrations of cracked Euler–Bernoulli beams. *Mech. Syst. Signal Proc.*, 68–69, 155–175.

9

Quantum-Inspired Topology Optimization

Topology optimization and quantum computing have evolved rapidly over the past three decades. Previous topological optimum design methods suffered from computational oppression and mathematical complexity. To overcome these shortcomings, a modified quantum-inspired evolutionary algorithm-based topology optimization method is proposed. This nested approach combines the quantum annealing strategy and the double-chains quantum genetic algorithm to establish an integral topology optimization framework. The former is used to determine the search direction of design variable updating without gradient information, while the latter ensures the abundant search diversity. The validity and feasibility of the developed methodology are demonstrated by several application examples. The results indicate that the proposed optimization framework is independent of initial values and can lead to optimized structures. It will also be more appropriate and effective if this strategy is deployed in a quantum computer in the future.

9.1. Introduction

With the evolution of additive manufacturing (AM) (Thompson *et al.* 2016) and other innovations in science and technology over the past three decades, structural topology optimization has become a useful design tool for obtaining efficient and lighter structures, since the pioneering work of Michell (1904) and the seminal work of Bendsøe and Kikuchi (1988). Topology optimization is the procedure of determining the connectivity, shape and location of voids inside a given design domain (Deaton and Grandhi 2014). It has proven to be a powerful technique for conceptual design, and is also of considerable practical interest for the fact that it can achieve far more significant savings than the conventional sizing or shape

Chapter written by Xiaojun WANG, Bowen NI and Lei WANG.

optimization (Rozvany 2001). In contrast to size and shape optimization, which involves variables such as thickness or cross-sectional areas of structural members and geometric variations pertaining merely to shape, topological optimization determines the overall topology/connectivity of the structural continuum and hence generates the optimum concept/configuration required in a design. Topology optimization is even regarded as the best method for solving structural optimal design problems and for producing the best overall structure (Tanskanen 2002).

Compared with other types of structural optimization, topology optimization of continuum structures is by far the most challenging technically and, at the same time, the most rewarding economically (Huang and Xie 2010). For topology optimization of continuum structures, the most common strategy in the literature is to initially discretize the allowable design space into finite elements (FE) and define the required loading/boundary conditions. The optimization procedure will then be mainly concerned with determining which elements should contain material (and so form the structure) and which elements are void (and thus represent the surrounding empty space). Various families of topology optimization methods have been well developed. The most established method is the one based on the homogenization approach first proposed by Bendsøe and Kikuchi (1988), while the evaluation of the optimization process is numerically complicated. Another essential alternative approach is the solid isotropic microstructure with penalization (SIMP) method (Rozvany *et al.* 1992; Bendsøe and Sigmund 1999), which has achieved reasonably general acceptance in recent years for its conceptual simplicity and computational efficiency (Bendsøe and Sigmund 2003). The evolutionary structural optimization (ESO) method is also a recognized method proposed by Xie and Steven (1993). Sigmund (2007) suggests that the researchers categorize ESO-type methods to the discrete form of the density-based approach for its similarities to the SIMP method; these two density-based approaches always encounter numerical difficulties such as mesh dependency, checkerboard patterns and local minima (Bendsøe and Sigmund 2003). Apart from density-based approaches, there exist several alternative approaches, such as the bubble method (Eschenauer *et al.* 1994), level-set method (Dijk *et al.* 2013), moving morphable component (MMC) method (Zhang *et al.* 2016) and phase-field method (Bourdin *et al.* 2003). Despite their distinct advantage in representing complex geometries, there is, however, still a long way to go before reaching the stage of regular industrial applications (Rozvany 2009). All the aforementioned families of methods cannot perform a global search and thus do not necessarily converge to the global optimal solution for the given objective function and constraints (Rozvany 2001; Zhou and Rozvany 2001). Furthermore, the final solution depends on the given initial design (Bendsøe and Sigmund 2003). Even under the same discretization and the same optimization strategy, different initial schemes may lead to different optimal results.

For many real-world problems, research works usually want to find the globally best solution in the feasible region because only the absolute best (extremum) is good enough (Neumaier 2004). As a result of this, more and more global optimization methods have been used in topology optimization, such as genetic algorithms (GAs) (Chapman *et al.* 1994; Jakiela *et al.* 2000; Wang and Tai 2005). Nevertheless, global optimization inevitably leads to much higher computational costs and financial oppression, especially in the circumstances of structural optimization problems, where function evaluation is a complete FE analysis in each iteration, making it an extremely challenging problem (Kane and Schoenauer 1997). Parallelization certainly brings a partial answer to this issue. However, it only distributes and does not reduce the CPU requirements needed for successful GA-based optimization problem-solving.

Coding is an effective way to transform design variables into those that are easily recognized and operated by programs. The most common and straightforward way to solve the topology optimization problem is 0-1 coding (Kane and Schoenauer 1997). The last three decades have seen the prosperous application of various properties from quantum physics to build quantum computers (Nielsen and Chuang 2011). Besides the “0” and “1” state in classic computers, quantum computers can manipulate a superposition of these two states. This characteristic is similar to the intermediate density element in the process of topological optimum design. In this sense, quantum-inspired evolutionary algorithms (QEA) are an exciting alternative to GAs, since quantum encoding of potential solutions considerably reduces the required number of chromosomes that guarantees good search diversity (Han and Kim 2002). In QEA, a solution individual is encoded by a quantum bit, which is expressed as a pair of normalized probability amplitude. Quantum bit coding represents 0–1 linear superposition and has abundant diversity (Glassner 2001). This enables QEA to obtain an ample search solution space even under a small population size, and get the global optimum with high probability. Various quantum gates such as the NOT gate, AND gate, OR gate, NAND gate, Hadamard gate and rotation gates can be applied to modify the state of a quantum bit (Hey 1998). The operation of the quantum gate update plays a role in updating the quantum chromosomes and makes quantum chromosomes evolve from generation to generation and converge to the optimal solution of the problem gradually.

In this chapter, a quantum-inspired evolutionary algorithm (QEA) for structural topology optimization is proposed. The rest of the chapter is organized as follows. In section 9.2, the formulation of the continuum structural topology optimization model is first reviewed, and the characteristics of quantum computing are also elucidated. In section 9.3, the detailed implementation of a QEA-based topology optimization is proposed, after a brief review of traditional SIMP-based topology optimization, and this is followed by a non-gradient quantum annealing (QA) strategy to enhance the effectiveness of the proposed scheme, in section 9.4. In section 9.5, two application cases are discussed to demonstrate the advantages of QEA-based topology optimization. Finally, some concluding remarks are provided in section 9.6.

9.2. General statements

In this section, a classical topology optimization model and some characteristics of quantum computing are reviewed. Two issues should be discussed: (1) the density-based continuum structural topology optimization formulation and (2) the characteristics of quantum computing, which may be conducive to reducing the computational expense of topology optimization (for details, see below).

9.2.1. *Density-based continuum structural topology optimization formulation*

The fundamental mathematical statement of a density-based topology optimization problem contains an objective function (which usually can be volume, compliance, displacements and stresses), a set of constraints (that is likely to be similar to the type of objective function) and a discretized representation of the physical system. Throughout this chapter, the mechanical model will be the standard two-dimensional plane stress linear model, and only linear static materials will be considered. The effects of gravity are also neglected. A general formulation based on linear static finite element analysis may be given as:

$$\begin{aligned} & \mathit{find}:\ \boldsymbol{\rho} = (\rho_1, \rho_2, \cdots, \rho_N) \\ & \min:\ f(\boldsymbol{\rho}, \mathbf{U}) \\ & s.t.\quad \mathbf{K}(\boldsymbol{\rho})\mathbf{U} = \mathbf{F} \\ & \qquad g_i(\boldsymbol{\rho}, \mathbf{U}) \le 0 \\ & \qquad 0 \le \rho_{\min} \le \boldsymbol{\rho} \le 1 \end{aligned} \qquad [9.1]$$

where $\boldsymbol{\rho}$ is the vector of density design variables, N is the number of design variables, f is the objective function and $\mathbf{U}$ and $\mathbf{F}$ are the displacement vector and the force vector, respectively. $\mathbf{K}$ is the global stiffness matrix, g_i are the constraints of the physical system and $\rho_{\min}$ is the lower bound of density variables.

In this chapter, we adopt the material interpolation model in the SIMP method, which has been widely used in recent years (Bendsøe and Sigmund 2003). To achieve a nearly solid-void design, Young's modulus of the intermediate material is interpolated as a function of the element density:

$$E(\rho_i) = E_1 \rho_i^p \qquad [9.2]$$

where E_1 denotes Young's modulus of the solid material and p depicts the penalty exponent. It is assumed that Poisson's ratio is independent of the design variables, and the global stiffness matrix $\mathbf{K}$ can be expressed by the summation of the elemental stiffness matrix and design variables ρ_i as:

$$\mathbf{K} = \sum_{i=1}^{N} \rho_i^p \mathbf{K}_i^0 \qquad [9.3]$$

where $\mathbf{K}_i^0$ denotes the elemental stiffness matrix of the ith solid element.

9.2.2. *Characteristics of quantum computing*

Over the past three decades, quantum computing, as an interdisciplinary science combining quantum theory with information science, has attracted extensive attention from many physicists and computer scientists, not only because it is an important way to apply quantum theory to solve practical problems, but also because it provides physicists with a profound insight into the principles of quantum mechanics. Compared with classical computation, the characteristic of the quantum computation process is discussed through the analysis of data storage, the calculation process and the output of results.

In conventional binary computing, the storage unit can only be represented by a one-bit binary number. Its basic state is "0" or "1", which is called a bit. When there are n such storage units, one n-bit data can be stored, representing a piece of information consisting of n-bit binary numbers, which is an n-bit register. Similarly, in quantum computation, the smallest unit of information is expressed by a qubit. The states in quantum mechanics have superposition, and the quantum bits can also be in superposition states of "0" and "1" states (Figure 9.1). The state of an n-qubit can be described as follows:

$$|\varphi_i\rangle = \alpha_i |0\rangle + \beta_i |1\rangle, i = 1, 2, \cdots, n \qquad [9.4]$$

where $|\cdot\rangle$ is the Dirac symbol of a quantum state and α_i and β_i are the probability amplitudes of the $|\varphi_i\rangle$ and satisfy the following normalization conditions:

$$|\alpha_i|^2 + |\beta_i|^2 = 1, i = 1, 2, \cdots, n \qquad [9.5]$$

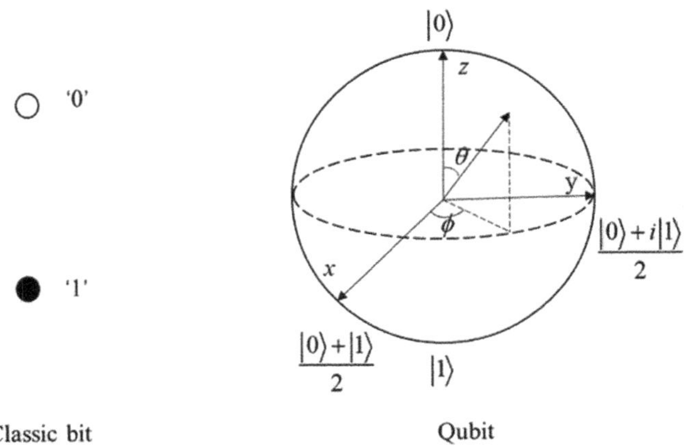

Figure 9.1. *Classic bit and qubit represented by two electronic levels in a sphere*

In n-qubit systems, there may be 2^n basic states: $|x_1 x_2 \cdots x_n\rangle$, where $x_i = 0$ or 1. n-qubit can be described as a linear combination of n orthogonal ground states:

$$|\varphi\rangle = \sum_{x \in \{0,1\}^n} a_x |x\rangle \qquad [9.6]$$

where a_x satisfies the normalization condition:

$$\sum_{x \in \{0,1\}^n} |a_x|^2 = 1 \qquad [9.7]$$

Therefore, an n-qubit quantum register can store 2^n n-bit binary numbers at the same time. The linear growth of the number of quantum registers therefore leads to the exponential growth of storage space, which is the basic feature of quantum computing memory cells and the premise that quantum computing speed can greatly exceed classical computing speed.

Corresponding to the logic gates of classical computing, the superposition states of quantum registers can also be transformed in quantum computing to achieve some logical functions. We will also touch on this in sections 9.3.2 and 9.3.3. A quantum computer performs one operation on an n-qubit quantum register, i.e. it performs mathematical operations on 2^n data stored at the same time, which is equivalent to repeating 2^n operations on a classical computer (Bennett and Divincenzo 1995). This is undoubtedly a significant improvement in computing speed.

We note that parallelism in classical computing is, in fact, a kind of incomplete parallelism. It does not change all the data stored by one operation, but only distributes the data to multiple processors to operate at the same time. Because of the superposition of quantum states, the quantum computation can realize the parallel computation in which all data can be changed by one operation completely.

9.3. Topology optimization design model based on quantum-inspired evolutionary algorithms

In this section, based on the SIMP method and the corresponding material interpolation model, a novel topology optimization approach, inspired by quantum computation and quantum-inspired evolutionary algorithms, is proposed in this section, in which the initial values of design variables will be replaced by chromosome populations. Moreover, additional solution aspects for the design variables updating in iteration process are systematically expounded, which may help to expand the search range of solution space and increase the possibility of finding the global optimal solution for the continuum structural topology optimization.

9.3.1. *Classic procedure of topology optimization based on the SIMP method and optimality criteria*

Stiffness is one of the key factors that must be taken into consideration in the design of structures such as aircrafts and bridges. Commonly, the compliance C ($C = \mathbf{F}^T\mathbf{U}$), the inverse measure of the overall stiffness of a structure, is considered. According to the topology optimization formulation and material interpolation scheme of a linear static system mentioned in section 9.2.1, if we take the compliance C into consideration and set it as the objective function and assume that the design variable $\boldsymbol{\rho}_i$ continuously changes from $\rho_{\min}$ to 1, the sensitivity of the objective function with respect to the change in the design variable is:

$$\frac{dC}{d\boldsymbol{\rho}_i} = \frac{d\mathbf{F}^T}{d\boldsymbol{\rho}_i}\mathbf{U} + \mathbf{F}^T \frac{d\mathbf{U}}{d\boldsymbol{\rho}_i} \qquad [9.8]$$

As described below, the adjoint method will be adopted to calculate the sensitivity of the displacement vector. By introducing a vector of Lagrangian multiplier $\boldsymbol{\lambda}$, an extra term $\boldsymbol{\lambda}^T(\mathbf{F}-\mathbf{KU})$ can be added to the objective function without changing anything due to the equilibrium equation $\mathbf{KU}=\mathbf{F}$. Thus

$$C = \mathbf{F}^T\mathbf{U} + \boldsymbol{\lambda}^T(\mathbf{F}-\mathbf{KU}) \qquad [9.9]$$

the sensitivity of the modified objective function can be written as

$$\begin{aligned}\frac{dC}{d\boldsymbol{\rho}_i} &= \frac{d\mathbf{F}^T}{d\boldsymbol{\rho}_i}\mathbf{U} + \mathbf{F}^T \frac{d\mathbf{U}}{d\boldsymbol{\rho}_i} + \frac{d\boldsymbol{\lambda}^T}{d\boldsymbol{\rho}_i}(\mathbf{F}-\mathbf{KU}) \\ &\quad + \boldsymbol{\lambda}^T\left(\frac{d\mathbf{F}}{d\boldsymbol{\rho}_i} - \frac{d\mathbf{K}}{d\boldsymbol{\rho}_i}\mathbf{U} - \mathbf{K}\frac{d\mathbf{U}}{d\boldsymbol{\rho}_i}\right)\end{aligned} \qquad [9.10]$$

Due to the equilibrium equation, the third term in the above equation is zero. In addition, it is assumed that the variation of an element does not affect the applied load vector, and therefore, $\frac{d\mathbf{F}}{d\boldsymbol{\rho}_i} = \mathbf{0}$. Thus, the sensitivity of the objective function becomes

$$\frac{dC}{d\boldsymbol{\rho}_i} = (\mathbf{F}^T - \boldsymbol{\lambda}^T\mathbf{K})\frac{d\mathbf{U}}{d\boldsymbol{\rho}_i} - \boldsymbol{\lambda}^T \frac{d\mathbf{K}}{d\boldsymbol{\rho}_i}\mathbf{U} \qquad [9.11]$$

The Lagrangian multiplier vector $\boldsymbol{\lambda}$ can be chosen freely based on equation [9.9]. The solution for $\boldsymbol{\lambda}$ is

$$\boldsymbol{\lambda} = \mathbf{U} \qquad [9.12]$$

by substituting $\boldsymbol{\lambda}$ into equation [9.11], the sensitivity of the objective function becomes

$$\frac{dC}{d\boldsymbol{\rho}_i} = -\mathbf{U}^T \frac{d\mathbf{K}}{d\boldsymbol{\rho}_i}\mathbf{U} \qquad [9.13]$$

by substituting the material interpolation scheme equation [9.3] into the above equation, the sensitivity of the objective function with regard to the change in the ith element can be found as

$$\frac{dC}{d\boldsymbol{\rho}_i} = -p\boldsymbol{\rho}_i^{p-1}\mathbf{U}_i\mathbf{K}_i^0\mathbf{U}_i \qquad [9.14]$$

In each iteration, after the sensitivities of all the elements pseudo-density $\boldsymbol{\rho}_i$ are obtained, the optimality criteria method is used for pseudo-density updating, as shown in equation [9.15]. For more details on the optimality criteria method, the reader is referred to the literature (Bendsøe 1995).

$$\rho_{new} = \begin{cases} \max(\rho_{\min}, \rho_i - \delta) & \text{if } \rho_i B_e^{\eta} \le \max(\rho_{\min}, \rho_e - \delta) \\ \rho_e B_e^{\eta} & \text{if } \max(\rho_{\min}, \rho_e - \delta) < \rho_e B_e^{\eta} < \min(1, \rho_e + \delta) \\ \min(1, \rho_e + \delta) & \text{if } \min(1, \rho_e + \delta) \le \rho_e B_e^{\eta} \end{cases} \qquad [9.15]$$

where δ denotes the step size of design variables updating, $B_e^{\eta} = \sqrt{\frac{-dC}{l_{mid}}}$, $\eta = \frac{1}{2}$ and l_{mid} represents the mid-value of the interval according to the dichotomy.

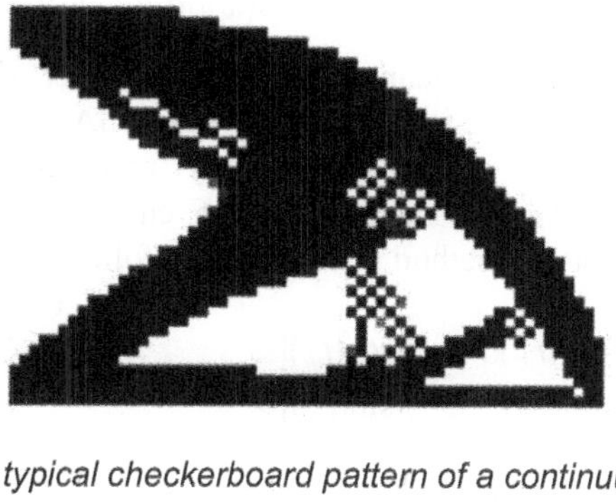

Figure 9.2. *A typical checkerboard pattern of a continuum structure*

When a continuum structure is discretized using low-order bilinear (2D) or trilinear (3D) finite elements, the sensitivity numbers could become C0 discontinuous across element boundaries. This leads to checkerboard patterns in the resulting topologies. Figure 9.2 shows a typical checkerboard pattern of a continuum structure from the original SIMP method. Furthermore, the mesh dependency, which refers to the problem of obtaining different topologies from different finite element meshes, is a neglectable shortcoming. In this chapter, the following sensitivity filter scheme (Sigmund and Petersson 1998) is adopted to overcome the above deficiencies:

$$\frac{\partial \hat{C}}{\partial \rho_e} = \frac{\sum_{i=1}^{n} \hat{H}_i \rho_i \frac{\partial C}{\partial \rho_i}}{\rho_e \sum_{i=1}^{n} \hat{H}_i} \qquad [9.16]$$

where $\hat{H}_i = r_{\min} - dist(e,i), \{i \in n, |dist(e,i) \le r_{\min}\}$, in which $\hat{H}_i$ is the convolution operator, $dist(e,i)$ denotes the distance of the center point of elements e and i, $r_{\min}$ is the filter radius, and n is the total number of the elements. The filter scheme requires little extra computational time and is very easy to implement in the optimization algorithm.

9.3.2. *The fundamental theory of a quantum-inspired evolutionary algorithm – DCQGA*

9.3.2.1. *Double chains encoding and decoding for quantum chromosomes*

Inspired by the concept of quantum computing, Li *et al.* (2011) proposed the double-chains quantum genetic algorithm (DCQGA). In the DCQGA, the probability amplitudes of qubits are directly regarded as the coding of a chromosome. Considering the randomicity of encoding and the restriction of equation [9.5], the encoding method is described as follows:

$$P_i = \begin{vmatrix} \cos(t_{i1}) & \cos(t_{i2}) & \cdots & \cos(t_{in}) \\ \sin(t_{i1}) & \sin(t_{i2}) & \cdots & \sin(t_{in}) \end{vmatrix} \qquad [9.17]$$

where $t_{ij} = 2\pi \times random\,[0,1]$, $i = 1,2,\cdots,m$, $j = 1,2,\cdots,n$, in which m represents the colony size, and n represents the number of qubits. In unit space $I^n = [-1,1]^n$, by equation [9.17], each chromosome simultaneously represents two solutions: the first line is the cosine solution and the other line is sine solution. Since two solutions are synchronously updated in each iteration step for the same colony size as the common genetic algorithm, such an encoding scheme can extend search range and accelerate the optimization process.

Each chromosome in the colony contains $2n$ probability amplitudes of n qubits. Each of the probability amplitudes corresponds to an optimization variable in solution space. If the jth qubit on the chromosome p_i is $[\alpha_i^j, \beta_i^j]^T$, then the corresponding pseudo-density design variables in solution space Ω can be computed as follows:

$$\begin{aligned}
&\boldsymbol{\rho}_{ic}^{j} = \frac{1}{2}\left[\left(1+\alpha_i^j\right)+\rho_{\min}\left(1-\alpha_i^j\right)\right] \\
&\boldsymbol{\rho}_{is}^{j} = \frac{1}{2}\left[\left(1+\beta_i^j\right)+\rho_{\min}(1-\beta_i^j)\right] \\
&i=1,2,\cdots,n, j=1,2,\cdots,m
\end{aligned} \quad [9.18]$$

Hence, each chromosome corresponds to two approximate solutions of the topology optimization problem.

9.3.2.2. *The rotation angle of quantum rotation gates*

Directed by the current optimal individual, the qubit chromosome updating process was performed by the quantum rotation gate defined as follows (Figure 9.3):

$$U(\Delta\theta)=\begin{bmatrix} cos(\Delta\theta) & -sin(\Delta\theta) \\ sin(\Delta\theta) & cos(\Delta\theta) \end{bmatrix} \quad [9.19]$$

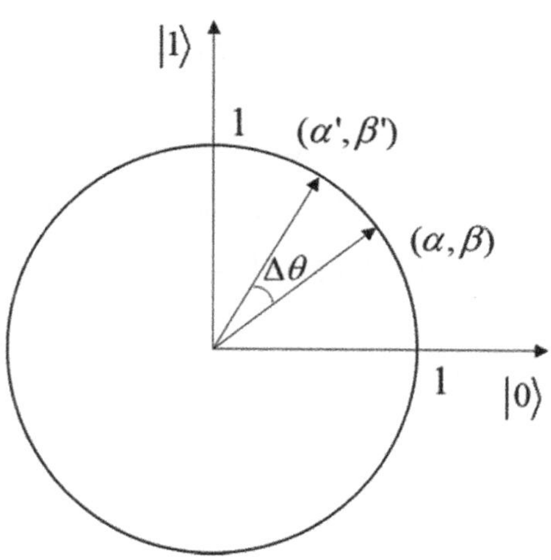

Figure 9.3. *Polar plot of the quantum gate for a qubit*

The update process can be expressed as:

$$\begin{bmatrix} cos(\Delta\theta) & -sin(\Delta\theta) \\ sin(\Delta\theta) & cos(\Delta\theta) \end{bmatrix} \begin{bmatrix} \cos t \\ \sin t \end{bmatrix} = \begin{bmatrix} \cos(t+\Delta\theta) \\ \sin(t+\Delta\theta) \end{bmatrix} \qquad [9.20]$$

where $\Delta\theta$ denotes the rotation angle. The magnitude and the direction of the rotation angle directly affect the convergence speed and direction, respectively. The method to determine the direction of the rotation angle is as follows: determine the probability amplitudes of the global optimum solution by far α_0, β_0 and the probability amplitudes of the current solution α_1, β_1, let

$$A = \begin{vmatrix} \alpha_0 & \alpha_1 \\ \beta_0 & \beta_1 \end{vmatrix} \qquad [9.21]$$

Then direction of the rotation angle $\Delta\theta$ can then be determined by such rules as follows: if $A \neq 0$, then the direction of $\Delta\theta$ is $-\text{sgn}(A)$, else, if $A = 0$, the direction of $\Delta\theta$ is arbitrary (both positive and negative directions will be accepted).

In this chapter, we adopt the following formula to calculate the magnitude of the rotation angle, based on the gradient method:

$$\Delta\theta_{ij} = -sgn(A)\Delta\theta_0 \exp\left(-\frac{\left|\nabla f\left(\mathbf{p}_i^j\right)\right| - \nabla f_j \min}{\nabla f_j \max - \nabla f_j \min}\right) \qquad [9.22]$$

where A is defined by equation [9.21], $\Delta\theta_0$ is the initial value of the rotation angle, usually defined as 0.001π, $\nabla f\left(\rho_i^j\right)$ is the gradient of the fitness function at $\mathbf{p}_i^j$, and $\nabla f_j \max$ and $\nabla f_j \min$ are, respectively, defined as follows:

$$\begin{aligned} \nabla f_j \max &= \max\left\{\left|\frac{\partial f(\mathbf{p}_1)}{\partial \mathbf{p}_1^j}\right|, \ldots, \left|\frac{\partial f(\mathbf{p}_m)}{\partial \mathbf{p}_m^j}\right|\right\} \\ \nabla f_j \min &= \min\left\{\left|\frac{\partial f(\mathbf{p}_1)}{\partial \mathbf{p}_1^j}\right|, \ldots, \left|\frac{\partial f(\mathbf{p}_m)}{\partial \mathbf{p}_m^j}\right|\right\} \end{aligned} \qquad [9.23]$$

where $\mathbf{p}_i^j$ represents the jth variable of vector $\mathbf{p}_i$ in solution space. If the current optimum solution is the cosine solution, then $\mathbf{p}_i^j = \mathbf{p}_{ic}^j$, else $\mathbf{p}_i^j = \mathbf{p}_{is}^j$. $\mathbf{p}_{ic}^j$ and $\mathbf{p}_{is}^j$ can be computed by equation [9.18], respectively.

We note that the DCQGA is not a quantum computer based algorithm, but rather, efficient intelligent algorithms for classical computers. In addition, the DCQGA is likely to present higher efficiency over traditional genetic algorithms when small populations are used. This feature is desirable in the continuum structural topology optimization because the use of large populations may well lead to prohibitive computational costs.

9.3.3. *Implementation of the integral topology optimization framework*

For general readers' ease of understanding, the logic and major steps of the presented methodology are successively summarized as follows.

According to the aforementioned work, the entire theoretical procedure of quantum-inspired topology optimization, which mainly combines FEM analysis, design sensitivity solution and quantum-inspired search, is accomplished for continuum structures.

i) **Input parameters**. Two kinds of data must be specified at the very beginning of the procedure: one type of data is the structural parameters, including material characteristics, size parameters, loads and boundary conditions; the other type is the optimization parameters, including the mutation rate, the crossover rate, the number of populations, the max generation and the convergence criterion.

ii) **Structural parameterization**. The proposed optimization framework uses a fixed-length vector to represent the design variables (see Figure 9.5). The topology variables are stored in an array and are encoded into integer numbers: 0 (void elements) and 1 (solid elements). The structural parameterization is performed in three steps. First, ρ_i are transformed from binary to decimal numbers ρ_i^d. Second, ρ_i^d are normalized:

$$\rho_i^{norm} = \frac{\rho_i^d}{2^n}(1-\rho_{\min})+\rho_{\min} \qquad [9.24]$$

where ρ_i^{norm} is the normalized value of each ρ_i^d, and n is the total number of qubits, which is also the number of design variables. $\rho_{\min}$ and 1 are the lower and the upper bounds, respectively, of the normalized interval. Finally, the numerical rounding is pursued and the ρ_i^{norm} collapse into different integer numbers (0 or 1) that will represent the encoded structure.

iii) **FEM analysis**. After discretizing the continuum structure using a fine mesh of finite elements, finite element analysis (FEA) is carried out for the structure. The objective function and constraints of topology optimization should be extracted from the results of FEA. This part can be realized by commercial finite element software such as ANSYS, ABAQUS and MSC.PATRAN/NASTRAN, or by programming tools such as MATLAB and PYTHON.

iv) **Design sensitivity solution**. First, the sensitivities of the objective function and constrains are generated. Then the sensitivities filter scheme mentioned in section 1.3.1 is pursued. Next, the optimality criterion (equation [9.15]) is adopted, to determine the direction of design variables' updating. Finally, the maximum variation of element pseudo-density in each iteration is stored as the objective function of the QEA.

v) **Quantum-inspired search**. First, the QEA performs local and global migration. If the best solution at the current generation B_i^{best} is better than the best overall solution b, then local migration is performed. This consists of replacing random individuals of B_i with the best-found solution. If the best solution at the current generation is not better than the previous best solutions, then global migration is applied, which consists of replacing all the individuals of B_i with b. Next, in each iteration, ρ_i is updated using the quantum rotation gate. Finally, if the convergence criterion is satisfied or the max evolution generation is achieved, finish the search process; the current best solution is the global optimum of the whole topology optimization procedure.

Figure 9.5 illustrates the flowchart for the presented QEA-based topology optimization methodology.

1	1	1	1	1	1	0	0	0	0
1	0	0	0	0	0	1	0	0	0
0	1	0	0	0	1	0	1	0	0
0	0	1	0	1	0	0	0	1	0
0	0	0	1	0	0	0	0	0	1
0	0	0	1	0	0	0	0	0	1
0	0	1	0	1	0	0	0	1	0
0	1	0	0	0	1	0	1	0	0
1	0	0	0	0	0	1	0	0	0
1	1	1	1	1	1	0	0	0	0

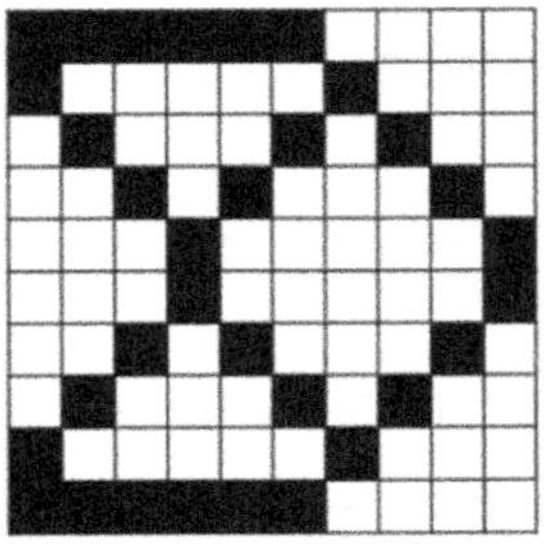

(a) Finite element encoding scheme (b) Topological form of the design area

Figure 9.4. *Finite element encoding scheme and topological form*

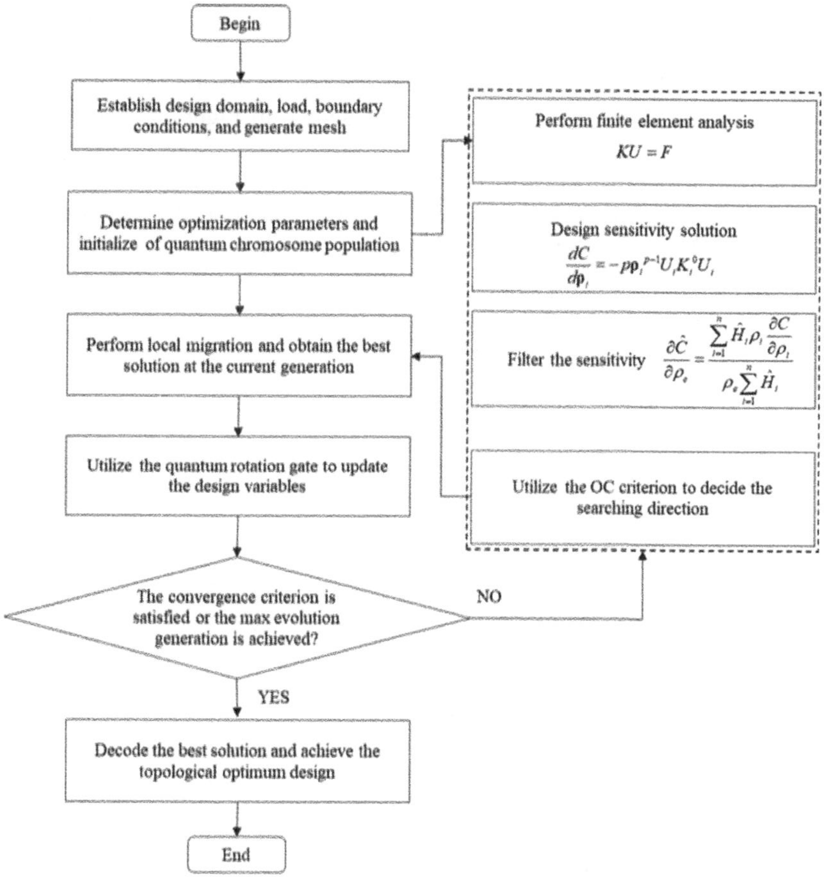

Figure 9.5. *Flowchart for the presented quantum-inspired methodology*

9.4. A quantum annealing operator to accelerate the calculation and jump out of local extremum

As everyone knows, numerous factors exist that affect the final result of genetic-like algorithms. During the specific implementation process of the above methodology, a non-gradient strategy is employed to replace the optimality criteria of the proposed topology optimization framework. A quantum annealing operator is applied to accelerate the calculation and jump out of the local extremum of the topological optimum design. More details are as follows.

Inspired by classic simulated annealing algorithm, quantum annealing is a promising optimization technique, which shows good performances on some typical optimization problems, such as the transverse Ising model and the traveling salesman problem (Martonák *et al.* 2004). As shown in Figure 9.6, instead of the "thermal jump" in classic simulated annealing, quantum annealing employs quantum fluctuations in frustrated systems or networks to anneal the system down to its ground state or its minimum cost state, tuning the quantum fluctuation eventually down to zero (Das and Chakrabarti 2005).

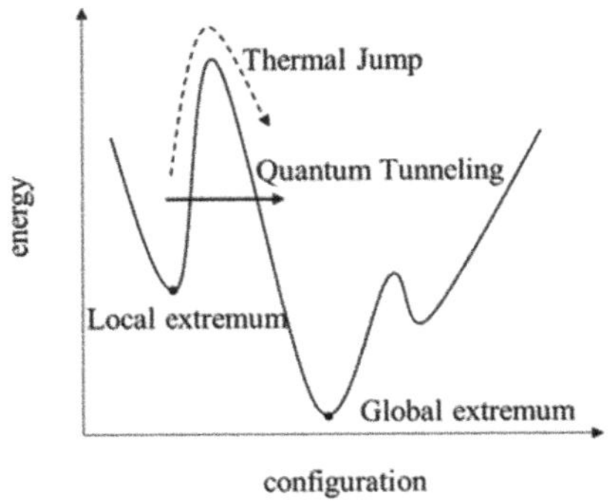

Figure 9.6. *Schematic indication of the advantage of quantum annealing over classical annealing*

Suppose that the topology optimization problem we wish to solve has been formulated as the minimization of a cost function, which we regard as a Hamiltonian H_0 of a classical many-body system without any external forces. Assume that $H'(t)$ denotes the result of external force. For the convenience of discussion, it is set to the transverse field $\Gamma(t)\sum_i \sigma_i^x$, then the Hamiltonian of the system can be expressed as:

$$H(t) = H_0 + H'(t) = H_0 + \Gamma(t)\sum_i \sigma_i^x \qquad [9.25]$$

where $\Gamma(t)$ represents the transverse field that causes the transition between different states, which is similar to the temperature T in simulated annealing. $\sum_i \sigma_i^x$ denotes the ith particle's Pauli representation in the x-axis. H_0 can be treated as the classical potential energy for a given configuration and $\Gamma(t)\sum_i \sigma_i^x$ as an appropriate kinetic energy that causes the necessary quantum transitions. If we set $H_0 = U$ and $\Gamma(t)\sum_i \sigma_i^x = K$, equation [9.25] can be written as:

$$H(t) = U + K \tag{9.26}$$

when the temperature is T (let Boltzmann's constant $k = 1$), the partition function (Martonák *et al.* 2004) of the system can be written as:

$$Z = Tr(e^{-\frac{H}{T}}) = Tr(e^{-\frac{H}{PT}})^P = Tr(e^{-\frac{U+K}{PT}})^P \tag{9.27}$$

where P is the number of particles in a certain state of the system and Tr is the summation in the quantum system. Therefore, the probability that the system is in the ith state is:

$$\rho_i = \frac{(e^{-\frac{H_i}{PT}})^P}{Z} \tag{9.28}$$

Similar to simulated annealing, the basic idea of quantum annealing is to regard the parameters as particles in the quantum system, and the objective function as the potential energy part of the Hamiltonian of the system. By slowly reducing the kinetic energy K (equivalent to the cooling process) of the system that promotes the quantum transition, iterative inversion is carried out to make the objective function finally reach the global minimum. In addition to the optimization criteria of simulated annealing, it also needs to meet the probability determined by the objective evaluation of the element itself, i.e. strain energy, and the larger the strain energy in the design domain the greater the probability of density change.

Here are the specific steps of performing the quantum annealing operator in the proposed procedure of quantum-inspired topology optimization. In each iteration of the framework in section 9.3.3:

i) **Define the design area**. Take the elements with better bearing capacity (the strain energy is close to the maximum value of the strain energy of the elements in the structure) and the elements with poor bearing capacity (the strain energy is close to the minimum value of the strain energy of the elements in the structure) as the areas to be optimized.

ii) **Initialize the parameters**. Calculate *delta* (the difference between the maximum and minimum strain energy of elements in the design area), $chance\_low$ (threshold of element density changes to "0") and $chance\_high$ (threshold of element density changes to "1").

$$\begin{aligned} delta &= \max_{sen} - \min_{sen} \\ chance\_low &= \frac{sen_i - \min_{sen}}{delta} \\ chance\_high &= \frac{\max_{sen} - sen_i}{delta} \end{aligned} \qquad [9.29]$$

where sen_i means the *i*th element strain energy. Then, the modified Metropolis rule can be written as:

$$p = \lambda * delta * T * e^{(-\frac{T}{loop} + loop)} \qquad [9.30]$$

where T is the temperature (its default value is 1000), loop denotes the iteration number and λ is the coefficient whose default value is 0.2.

iii) **Update the design variables**. For every element in the design area which meets the $chance\_low$ or $chance\_high$ threshold conditions, if $p > rand$ ($rand$ represents a random number between 0 and 1), then keep the element density, or update the element density to "0" or "1", as shown in Figure 9.7.

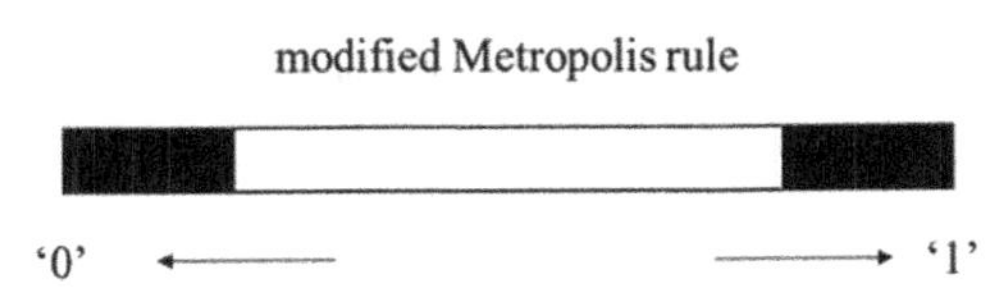

Figure 9.7. *The diagram of element density updating*

The proposed quantum annealing operator is an alternative approach to replace optimality criteria and sensitivity analysis, which can be verified by the numerical example in section 9.5.2. It gives excellent performance during the process of quantum-inspired topology optimization. It is worth noting that this operator can be applied to a classic topology optimization framework independently, which also helps to expand the search range of solution space.

9.5. Numerical examples

In this section, a planar numerical case is presented to validate the rationality and superiority of the developed QEA-based topology optimization methodology, for continuum structure design problems. To clarify the feasibility and effectiveness of the presented optimization framework in practical engineering, a more complicated wing rib case is further analyzed. In the examples, the four-node plane stress element is used for finite element analysis.

9.5.1. *Example of a short cantilever*

The short cantilever shown in Figure 9.8 is under plane stress conditions and a vertical load of 1 is applied in the middle of the free end. Young's modulus $E = 1$ and Poisson's ratio $v = 0.3$ are assumed. The design domain is divided into 60×40 four-node elements. For this simulation, the population size is set at 100 and the maximum number of generations is set at 500. The rotation increment $\Delta\theta_0$ is set at 0.001π, and the mutation probability p_m is set at 0.2.

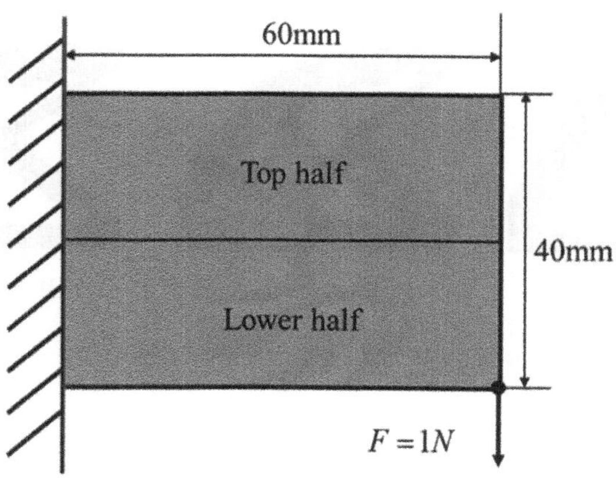

Figure 9.8. *Design domain of a short cantilever*

In Figure 9.9, we refer to the 99 line code proposed by Sigmund (2001), and set a group of different initial values in the classic SIMP method. The different choices of the initial distribution of material schemes will lead to various topological results. Worse still, it is difficult to distinguish which structure is the best.

(a) Top 0.1, lower 0.9 (b) Top 0.9, lower 0.1 (c) Top 0.5, lower 0.5

Figure 9.9. *Result of different initial distribution of material schemes*

Figure 9.10 illustrates the result of three different topology optimization methods. The left, the middle and the right are the classic SIMP method (initial density value is 1), the GA-based method and the QEA-based method, respectively. We note that while the compliance is quite similar, the design under QEA resembles a more conventional truss structure than the SIMP method. The optimal layout has avoided sharp angles and has fewer trusses that give rise to stress concentrations than the GA-based method.

(a) SIMP method (b) GA-based method (c) QEA-based method

Figure 9.10. *Result of different methods*

9.5.2. *Example of a wing rib*

In this example, a wing rib structure with some weight-reducing holes is clamped in the left and multi loads are applied at the points of the upper and lower edge strips, $F_{1x} = F_{1y} = 1N$ and $F_{2x} = F_{2y} = -1N$, as shown in Figure 9.12. Young's modulus E = 1 MPa and Poisson's ratio ν = 0.3 are assumed. The minimum structural compliance of the wing rib is taken as the optimization objective, and the volume fraction constraint is set to 0.45. In this case, both the SIMP method and independent QA-based strategy are carried out. The final results are shown in Figure 9.12, respectively. In QA-based strategy, the initial configuration is achieved by two iterations of design variables updating with optimality criteria. The number of iterations and the final compliance of the two methods are shown in Table 9.1.

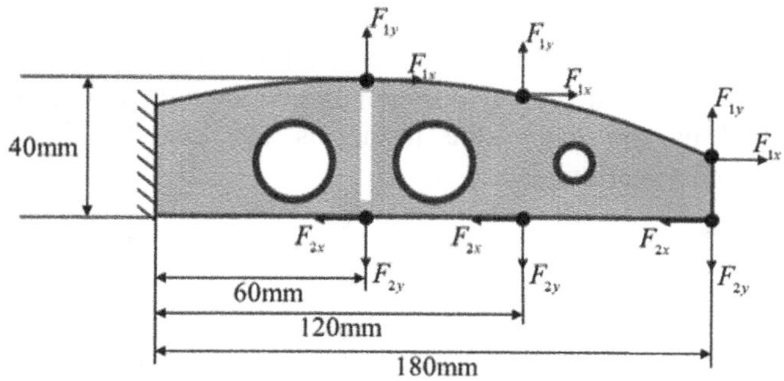

Figure 9.11. *A wing rib structure under multiple loads*

(a) SIMP method (b) QA-based strategy

Figure 9.12. *Topological configuration of wing rib under different methods*

	SIMP method	QA-based strategy
Number of iterations	82	100
Final compliance ($N \cdot mm$)	281.752	276.886

Table 9.1. *The number of iterations and final compliance of different methods*

From the results given above, the following points can be concluded:

1) In the topology optimization of this wing rib, both methods can achieve an optimal material layout, which satisfies the volume fraction constraint.

2) As indicated by the black circle, compared to the optimal layout originating from the SIMP method, the result from the QA-based strategy has fewer trusses that give rise to stress concentrations. From the perspective of processing technology, the material layout of the edge strip is more uniform with the QA strategy; therefore, it is more convenient for manufacturing.

3) It can be seen from the results of these two methods (in Figure 9.13 and Table 9.1) that under the same volume fraction constraint, the compliance of the wing rib with SIMP method is 281.752 *N mm*, and the compliance of the wing rib

with QA-based strategy is 276.886, which means that the optimal wing rib structure has greater rigidity and can endure a more critical load. Besides, the number of iterations with QA-based strategy does not exceed the number of iterations with the SIMP method by too much (the former is 100, and the latter is 82). The computational cost is affordable.

9.6. Conclusion

In this article, an alternative quantum-inspired optimization framework applied to the topology optimization of continuum structures was proposed. The method combines the classic SIMP model and QEA algorithms (in this chapter, we adopted the DCQGA and QA algorithms). Based on the obtained results, the following conclusion can be drawn.

When performing the topology optimization procedure, the increase in the continuum structure discretization led to a sharp increase in the search space. For the traditional heuristic algorithms, it is bound to bring about a sharp increase in computational effort. In this sense, quantum-inspired algorithms proved an interesting stochastic search method for the optimization of complicated continuum structures, with excellent performance in expanding the search range of the solution space. A QA-based strategy is proposed to accelerate the update of the design variable. It is a non-gradient, effective and reliable alternative for the traditional gradient-based optimality criteria because it does not need to perform sensitivity analysis and increase the number of iterations too much.

Nevertheless, these good results must not hide the main drawback of the proposed method, namely, its computational cost, which makes it highly unlikely to be applicable to real-world optimization works. Furthermore, as the experimental implementation of quantum computation needs initialization, coherent manipulation, control and read of the fragile quantum system, practically building quantum computers has proved extremely difficult. However, we still believe that in the near future, the advantages of quantum computing and quantum-inspired algorithms in topology optimization will be fully realized.

9.7. Acknowledgments

The work of this paper is supported by the National Nature Science Foundation of China (No.12072006, No.11872089) and the Defense Industrial Technology Development Program (Nos. JCKY2017601B001, JCKY2017208B001, JCKY2018601B001, JCKY2019203A003, JCKY2019209C004).

9.8. References

Bendsøe, M.P. (1995). *Optimization of Structural Topology, Shape, and Material*. Springer, Berlin.

Bendsøe, M.P. and Kikuchi, N. (1988). Generating optimal topologies in structural design using a homogenization method. *Computer Methods in Applied Mechanics & Engineering*, 71(2), 197–224.

Bendsøe, M.P. and Sigmund, O. (1999). Material interpolation schemes in topology optimization. *Archive of Applied Mechanics*, 69(9–10), 635–654.

Bendsøe, M.P. and Sigmund, O. (2003). *Topology Optimization: Theory, Methods, and Applications*. Springer, Berlin.

Bennett, C.H. and Divincenzo, D.P. (1995). Quantum information and computation. *Nature*, 48(10), 24–30.

Bourdin, B. and Chambolle, A. (2003). Design-dependent loads in topology optimization. *Esaim Control Optimisation & Calculus of Variations*, 9(9), 19–48.

Chapman, C.D., Saitou, K., Jakiela, M.J., Noyce, R.N. (1994). Genetic algorithms as an approach to configuration and topology design. *ASME Journal of Mechanical Design*, 116(1005).

Das, A. and Chakrabarti, B.K. (eds) (2005). *Quantum Annealing and Related Optimization Methods*. Springer. Berlin.

Deaton, J.D. and Grandhi, R.V. (2014). A survey of structural and multidisciplinary continuum topology optimization: Post 2000. *Structural & Multidisciplinary Optimization*, 49(1), 1–38.

Dijk, N.P.V., Maute, K., Langelaar, M., Keulen, F.V. (2013). Level-set methods for structural topology optimization: A review. *Structural & Multidisciplinary Optimization*, 48(3), 437–472.

Eschenauer, H.A., Kobelev, V.V., Schumacher, A. (1994). Bubble method for topology and shape optimization of structures. *Structural Optimization*, 8(1), 42–51.

Glassner, A.S. (2001). Quantum computing, part 3. *IEEE Computer Graphics & Applications*, 21(4), 84–92.

Han, K.-H. and Kim, J.-H. (2002). Quantum-inspired evolutionary algorithm for a class of combinatorial optimization. *IEEE Transactions on Evolutionary Computation*, 6(6), 580–593.

Hey, T. (1998). Quantum computing: An introduction. *Computing & Control Engineering Journal*, 10(3), 105–112.

Huang, X.D. and Xie, Y.M. (2010). *Evolutionary Topology Optimization of Continuum Structures: Methods and Applications*. Wiley, Hoboken.

Jakiela, M.J., Chapman, C., Duda, J., Adewuya, A., Saitou, K. (2000). Continuum structural topology design with genetic algorithms. *Computer Methods in Applied Mechanics & Engineering*, 186(2), 339–356.

Kane, C. and Schoenauer, M. (1997). Topological optimum design using genetic algorithms. *Control and Cybernetics*, 25(5), 1059–1088.

Li, P.C., Song, K.P., Shang, F.H. (2011). Double chains quantum genetic algorithm with application to neuro-fuzzy controller design. *Advances in Engineering Software*, 42(10), 875–886.

Martonák, R., Santoro, G., Tosatti, E. (2004). Quantum annealing of the traveling salesman problem. *Physical Review E: Statistical, Nonlinear, and Soft Matter Physics*, 70, 057701.

Michell, A.G.M. (1904). The limits of economy of materials in frame structures. *Philosophical Magazine*, 6(8), 589–597.

Neumaier, A. (2004). Complete search in continuous global optimization and constraint satisfaction. *Acta Numerica*, 13(1), 271–369.

Nielsen, M.A. and Chuang, I.L. (2011). *Quantum Computation and Quantum Information*, 10th anniversary edition. Cambridge University Press, Cambridge.

Rozvany, G.I.N. (2001). Aims, scope, methods, history and unified terminology of computer-aided topology optimization in structural mechanics. *Structural & Multidisciplinary Optimization*, 21(2), 90–108.

Rozvany, G.I.N. (2009). A critical review of established methods of structural topology optimization. *Structural & Multidisciplinary Optimization*, 37(3), 217–237.

Rozvany, G.I.N., Zhou, M., Birker, T. (1992). Generalized shape optimization without homogenization. *Structural Optimization*, 4(3–4), 250–252.

Sigmund, O. (2001). A 99 line topology optimization code written in Matlab. *Structural & Multidisciplinary Optimization*, 21(2), 120–127.

Sigmund, O. (2007). Morphology-based black and white filters for topology optimization. *Structural & Multidisciplinary Optimization*, 33(4–5), 401–424.

Sigmund, O. and Petersson, J. (1998). Numerical instabilities in topology optimization: A survey on procedures dealing with checkerboards, mesh-dependencies and local minima. *Structural Optimization*, 16(1), 68–75.

Tanskanen, P. (2002). The evolutionary structural optimization method: Theoretical aspects. *Computer Methods in Applied Mechanics & Engineering*, 191(47), 5485–5498.

Thompson, M.K., Moroni, G., Vaneker, T., Fadel, G., Campbell, R.I., Gibson, I., Ahuja, B. (2016). Design for additive manufacturing: Trends, opportunities, considerations, and constraints. *CIRP Annals – Manufacturing Technology*, 65(2), 737–760.

Wang, S.Y. and Tai, K. (2005). Structural topology design optimization using genetic algorithms with a bit-array representation. *Computer Methods in Applied Mechanics & Engineering*, 194(36), 3749–3770.

Xie, Y.M. and Steven, G.P. (1993). A simple evolutionary procedure for structural optimization. *Computers & Structures*, 49(5), 885–896.

Zhang, W., Li, D., Yuan, J., Song, J., Xu, G. (2016). A new three-dimensional topology optimization method based on moving morphable components (MMCs). *Computational Mechanics*, 59(4), 1–19.

Zhou, M. and Rozvany, G.I.N. (2001). On the validity of ESO type methods in topology optimization. *Structural & Multidisciplinary Optimization*, 21(1), 80–83.

10

Time Delay Vibrations and Almost Sure Stability in Vehicle Dynamics

Road vehicle dynamics concern multi-body car models rolling on an uneven road with random profiles. Because of the front and rear wheels, road excitations affect time delay vibrations with resonance and absorption at countably many speeds of the car. The applied Gaussian road excitation is extended to stochastic sinusoidal models with bounded realizations, assuming that the car speed is additively perturbed by noise. This chapter introduces random amplitude processes of quarter-car models and investigates their resonance and stability behavior.

10.1. Introduction to road vehicle dynamics

In order to introduce the basics of road vehicle dynamics, Figure 10.1 shows a quarter-car model that rolls on a wavy road with amplitude Z and frequency Ω given by the wavelength $L = 2\pi/\Omega$. In the stationary case, when the car is driving with constant speed v, vertical car vibrations excited by the wavy road are described by the ratio Y/Z of the response amplitude Y to the excitation amplitude Z. This amplitude ratio (see, for example, Den Hartog 1934 and Klotter 1978) is plotted in Figure 10.2 for two damping values D against $\nu = \mathrm{v}\,\Omega/\omega_1$, where the speed frequency $\nu\Omega$ is related to the natural frequency ω_1 of the car. In Figure 10.2, λ is an image variable of the speed axis with two scales: $\lambda = \nu$ on the left half of the figure and $\lambda = 2 - 1/\nu$ on the right half of the figure. This scaling (see, for example, Klotter 1955) has the advantage that amplitude ratios can be shown in the whole speed range $0 \leq \nu < \infty$. Well-known results are presented in Figure 10.2, which are as follows: when driving slowly, cars follow the road contour identically without

Chapter written by Walter V. WEDIG.

any magnification of the relative ground motion. When driving with the resonance speed $v = 1$, the vertical vibrations become maximal and vanish completely when the car speeds up to an infinitely high velocity. More realistic road profiles are random processes (see, for example, Elishakoff 1983). They are normally distributed with zero mean and root mean square (rms), or standard deviation σ_z, which that replaces the response amplitude in the deterministic case.

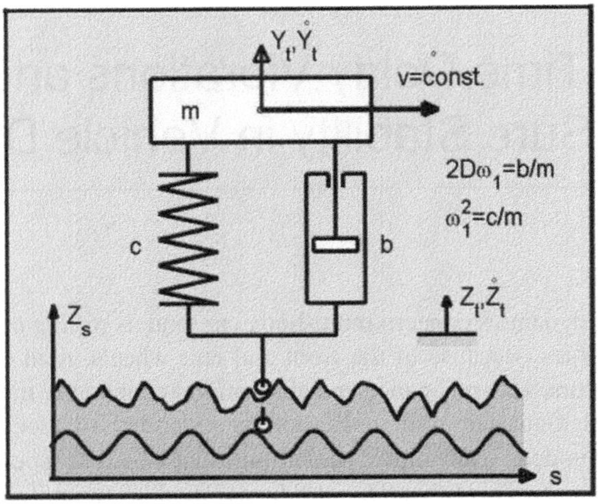

Figure 10.1. *Quarter-car model rolling at constant speed on an uneven road with a sinusoidal or stochastic level profile. For a color version of this figure, see www.iste.co.uk/challamel/mechanics3.zip*

The one-dimensional road processes applied in the first part of this chapter are characterized by means of first-order spectra, defined by Doods and Robson (1973) (also see Sobczyk *et al.* (1977) and Davis and Thompson (2001)).

The calculated rms value of vertical car vibrations, related to that of road excitations, is plotted in Figure 10.2 for the same damping values D against the related speed frequency $v = \mathrm{v}\,\Omega/\omega_1$, where Ω denotes the corner frequency of the road spectra (see Wedig 2003). Obviously, resonance amplifications are much stronger in the deterministic case than in the stochastic case when the car is driving in the resonance range around $v = 1$. Outside the resonance, this situation is inverted completely. The rms ratio (blue) of the car response and the road excitation is much larger than the amplitude ratio (red) in the harmonic case, particularly when the car is driving with higher speed in the overcritical range. However, in both cases, the magnification ratios coincide when the car slows down to zero speed or speeds up to an infinitely high speed.

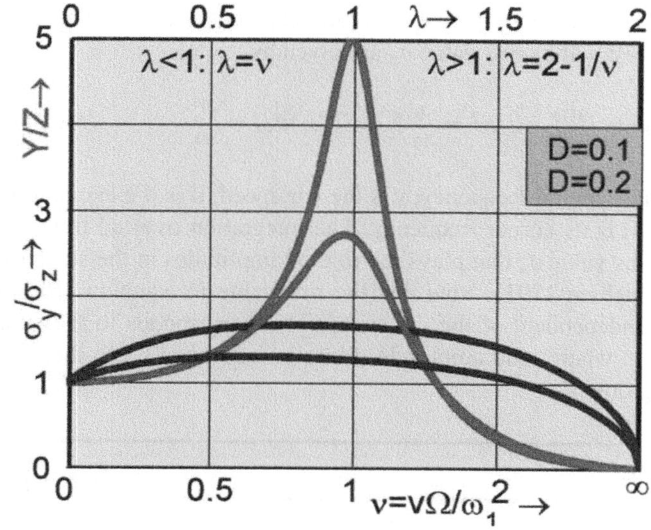

Figure 10.2. *Amplitude ratio (red) and rms ratio (blue) for two damping values against the related speed frequency. For a color version of this figure, see www.iste.co.uk/challamel/mechanics3.zip*

10.2. Delay resonances of half-car models on road

Figure 10.3 extends the quarter-car model to a half-car model rolling on a random road surface that affects planar car body motions with two degrees of freedom. The angle and vertical vibrations are described by two equations of motion. In the symmetric case, both equations of motion are decoupled and can be written as

$$\ddot{X}_t + 2D\omega_1\dot{X}_t + \omega_1^2 X_t = D\omega_1\dot{U}_t + \omega_1^2\, U_t/2 \qquad [10.1]$$

$$U_t = Z_{t+T/2} \pm Z_{t-T/2},\ T = a/\mathrm{v} \qquad [10.2]$$

where indexed capital letters denote set functions that are dependent on time t.

The positive sign in the delay equation [10.2] denotes the vertical vibration X_t, and the negative sign denotes the angle vibrations. The parameters ω_1 and D given in equation [10.1] of motion determine the natural frequency of the car and its damping, respectively. Both parameters are given by $\omega_1^2 = c/m$ and $2D\omega_1 = b/m$ with mass m, spring c and damper b. There are two ground excitations: one at the front wheel and the other at the rear wheel. Both excitations are delayed (see Di Poala and Pirotta 2001) by the time difference $T = a/v$, given by the wheelbase a divided by the constant car speed $\mathrm{v} \geq 0$. The road spectrum $S_z(\omega)$ of interest and its rms value σ_z are given by

$$S_z(\omega) = \frac{\sigma^2 \mathrm{v}}{\omega^2+(\mathrm{v}\Omega)^2}, \quad \sigma_z^2 = \frac{1}{2\pi}\int_{-\infty}^{+\infty} S_z(\omega) d\omega = \frac{\sigma^2}{2\Omega} \qquad [10.3]$$

where ω is the spectral frequency, v is the car speed, σ is the excitation intensity of the road and Ω is its corner frequency. The integration over all time frequencies ω leads to the rms value σ_z that plays the role of amplitudes in the stochastic case (see Popp and Schiehlen 1993). Note that the rms value in equation [10.3] of the road excitation is independent of the car speed. This corresponds to the situation in the harmonic case, where road amplitudes remain unchanged when the car is speeding up or slowing down.

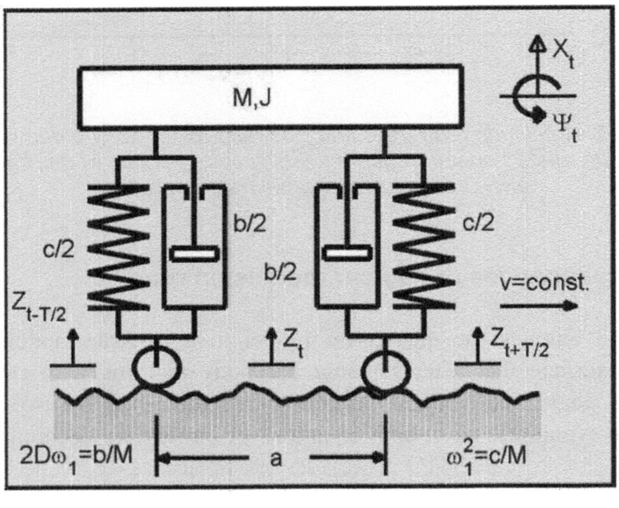

Figure 10.3. *Half-car model rolling on a random road profile with two degrees of freedom and the center distance* ***a*** *between the front and rear wheels. For a color version of this figure, see www.iste.co.uk/challamel/mechanics3.zip*

The delay terms in equation [10.2] lead to the spectrum $S_u(\omega) = 1 \pm \cos\omega T$, which needs to be multiplied by the frequency responses of the equation of motion [10.1] and its conjugate complex version, in order to obtain the car response spectrum

$$S_x(\omega) = \frac{[\omega_1^4+(2D\omega_1\omega)^2](1\pm\cos\omega T)}{2[(\omega_1^2-\omega^2)^2+(2D\omega_1\omega)^2]} S_z(\omega) \qquad [10.4]$$

For $D \leq 1$ and the + sign in equation [10.4], the integration over all frequencies ω leads to the squared standard deviation or mean square ratio of the response and excitation:

$$\frac{\sigma_x^2}{\sigma_z^2} = \frac{1}{2}\{ [1-(2D\nu)^2] + \frac{\nu}{2D}[1+\nu^2+(2D\nu)^2]\left[e^{-\frac{D\alpha}{\nu}}\cos\left(\frac{\alpha}{\nu}\sqrt{1-D^2}\right)+1\right] \quad [10.5]$$

$$+\frac{\nu}{2\sqrt{1-D^2}}[3+\nu^2-(2D\nu)^2]\,e^{-\frac{D\alpha}{\nu}}\sin\left(\frac{\alpha}{\nu}\sqrt{1-D^2}\right)\}\,\frac{1}{(\nu^2+1)^2-(2D\nu)^2}$$

where $\nu = \mathrm{v}\Omega/\omega_1$ is the related speed frequency of the car and $\alpha = a\Omega$ is its dimensionless wheelbase. For the special case $\alpha = 0$ of the quarter-car model, the lower part in equation [10.5] vanishes. Only the upper part remains and simplifies to

$$\frac{\sigma_x^2}{\sigma_z^2} = \frac{1}{2D}\,\frac{2D+(1+4D^2)\nu}{1+2D\nu+\nu^2}, \quad \nu = \mathrm{v}\Omega/\omega_1 \qquad [10.6]$$

The same result is obtained, but multiplied by the factor $1/2$, for the limiting case $\alpha \to \infty$, in which the wheelbase is infinitely large. In this case, the ground excitations through the front and rear wheels are statistically independent. The results of $\alpha = 0$ and $\alpha = \infty$ are evaluated in Figure 10.4, which are shown by the black lines. The upper black curve is already shown in Figure 10.4, applying a different scaling for the vertical axis.

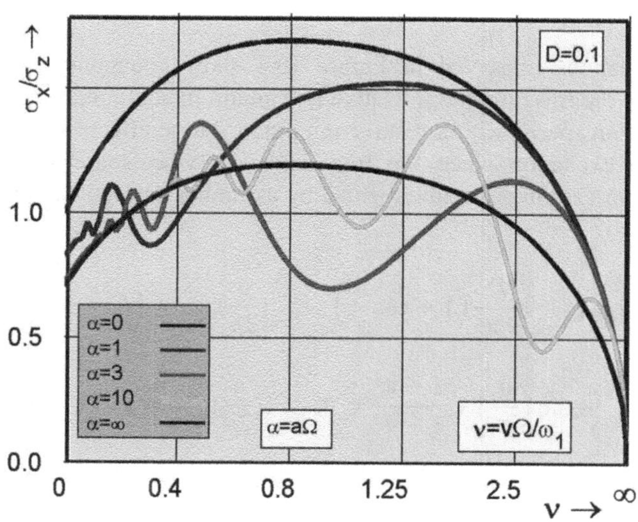

Figure 10.4. *Root mean square ratios of the response and excitation of the half-car model for* $\mathbf{D = 0.1}$ *and for five different wheel distances. For a color version of this figure, see www.iste.co.uk/challamel/mechanics3.zip*

In Figure 10.4, the rms ratio [10.5] is plotted against the related speed frequency for the damping $D = 0.1$ and the wheelbase $\alpha = 1, 3, 10$, represented by blue, red and green, respectively.

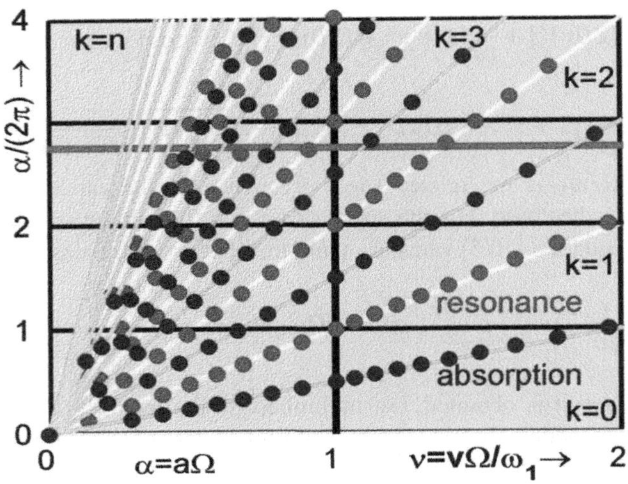

Figure 10.5. *Resonance and absorption speeds for k=1,2,3,... calculated from equation [10.5] for D<<1 and wheelbases $\alpha \leq 8\pi$. For a color version of this figure, see www.iste.co.uk/challamel/mechanics3.zip*

The evaluations presented in Figure 10.4 show countably many resonance speeds, where the rms ratio has a relative maximum. In addition, there are countably many absorption speeds with a relative minimum. These effects are due to the time delay in the excitation when the front and rear wheels are separated. In the following, both extremes are investigated by assuming that the damping D tends to zero, leading to

$$\left(\frac{\sigma_x}{\sigma_z}\right)^2 |_{D\ll 1} = \frac{\nu/D}{1+\nu^2}\left(1+\cos\frac{\alpha}{\nu}\right), \qquad \left(\frac{\alpha}{2\pi}\right)_{min} = \left(k - \frac{1}{2}\right)\nu$$

$$\frac{\alpha}{\nu}\tan\left(\frac{\alpha}{2\nu}\right) + \frac{1-\nu^2}{1+\nu^2} = 0, \qquad \left(\frac{\alpha}{2\pi}\right)_{max} = k\nu,\ k = 1,2,3\ ...$$

which is differentiated with respect to ν to obtain the resonances (yellow–red) and absorptions (cyan–blue) for all $\alpha/(2\pi) \leq 4$ and $\nu \leq 2$, as shown in Figure 10.5.

10.3. Extensions to multi-body vehicles on a random road

According to Wedig (2003), realizations of random road profiles are assumed to be normally distributed, generated, for example, by means of the first-order system:

$$\dot{Z}_t = -\mathrm{v}\Omega Z_t + \sigma\sqrt{\mathrm{v}}\xi_t,\; E(\xi_t\xi_{t+\tau}) = \delta(\tau) \qquad [10.7]$$

which is subjected to Gaussian white noise ξ_t with zero mean, delta correlation and intensity σ. Its power spectrum is uniformly distributed over all frequencies ω, i.e. $S_\xi(\omega) = 1$. Consequently, the application of the Fourier transform in equation [10.7] multiplied by its conjugate version leads to the road spectrum $S_z(\omega)$, which is already noted in equation [10.3]. The parameter v denotes the velocity with which the car is rolling on the road. The speed becomes negative when the car is driving backwards. In this case, the speed parameter v must be replaced by its absolute value in order to ensure existence and stability in equation [10.7] (see Wedig 2012; 2016b). The road model is applied to the multi-body vehicle, as shown in Figure 10.6. It has two viscoelastically supported wheels m and one car body with mass M and moment of inertia $J = k_s^2 M$. Dampers and springs are denoted by b, c and k, respectively. Two reference frequencies are introduced with $\omega_1^2 = 2\,k/M$ and $\omega_2^2 = c/m$ given by $\omega_1 = 1/s$ and $\omega_2 = 10/s$. The damping is given by $2D\omega_1 = 2\,b/M$ with the value $D = 0.1$. The parameters $\mu = 4$ and $\rho = 2$ denote, respectively, the mass ratio $\mu = M/(2m)$ and the inertia radius $\varrho = a/(2k_s)$, where a is the wheelbase applied by $\alpha = a\Omega = 10$. The coordinates x_1 and x_2 describe the vertical wheel vibrations. The vertical motion of the car body is described by x and $y = \phi\, a/2$. Because of the two wheels, the road excitations at the front and rear wheels are delayed by $T = a/\mathrm{v}$, where v is the car speed that is assumed to be constant. Both excitations are generated by the two road equations

$$\dot{Z}_t = -\mathrm{v}\Omega Z_t + \sigma\sqrt{\mathrm{v}}\xi_{t+T/2},\; \dot{U}_t = -\mathrm{v}\Omega U_t + \sigma\sqrt{\mathrm{v}}\xi_{t-T/2}$$

where the time delay T is introduced into the white noise term by $\pm T/2$.

The four equations of motion of the half-car model are rewritten into eight first-order equations. Together with two excitations, a tenth-order system is obtained by

$$\dot{\vec{V}}_t = A\vec{V}_t + \vec{\sigma}_p\xi_{t+T/2} + \vec{\sigma}_m\xi_{t-T/2},\; \vec{V}_t = \left(X_t\dot{X}_tY_t\dot{Y}_tX_{1t}\dot{X}_{1t}X_{2t}\dot{X}_{2t}Z_tU_t\right)'$$

$$\vec{\sigma}_p = \sigma v(\,0000\,|0000\,|01\,)',\; \vec{\sigma}_m = \sigma v(\,0000\,|0000\,|10)'$$

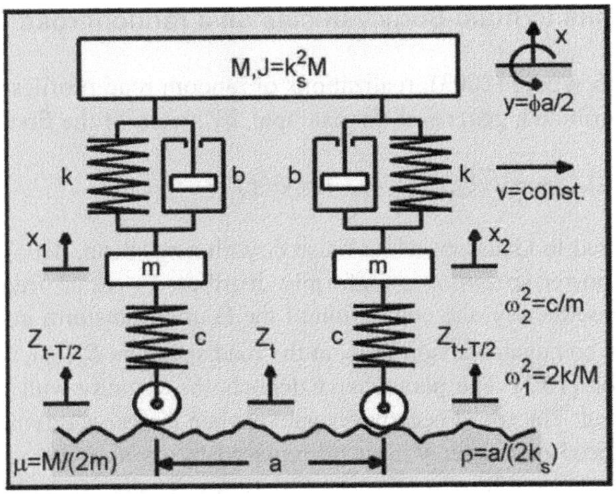

Figure 10.6. *Extended half-car model with four degrees of freedom and two viscoelastically supported wheels rolling on a random road. For a color version of this figure, see www.iste.co.uk/challamel/mechanics3.zip*

The above vector equation is integrated and multiplied by its transposed form to obtain the covariance equation, by means of the expectation operator in the matrix form

$$E(\vec{V}_t\vec{V}_t') = \int_{-\infty}^{t} \mathrm{e}^{A(t-\tau)}\left[\vec{\sigma}_\mathrm{p}\vec{\sigma}_\mathrm{p}' + \vec{\sigma}_\mathrm{m}\vec{\sigma}_\mathrm{m}' + \mathrm{e}^{\mathrm{AT}}\vec{\sigma}_\mathrm{p}\vec{\sigma}_\mathrm{m}' + \vec{\sigma}_\mathrm{m}\vec{\sigma}_\mathrm{p}'\mathrm{e}^{\mathrm{A'T}}\right]\mathrm{e}^{A'(t-\tau)}\mathrm{d}\tau.$$

This result is differentiated with respect to the time t to obtain the stationary form

$$AE(\vec{V}_t\vec{V}_t') + E(\vec{V}_t\vec{V}_t')A' + \vec{\sigma}_\mathrm{p}\vec{\sigma}_\mathrm{p}' + \vec{\sigma}_\mathrm{m}\vec{\sigma}_\mathrm{m}' + \mathrm{e}^{\mathrm{AT}}\vec{\sigma}_\mathrm{p}\vec{\sigma}_\mathrm{m}' + \vec{\sigma}_\mathrm{m}\vec{\sigma}_\mathrm{p}'\mathrm{e}^{\mathrm{A'T}} = 0 \qquad [10.8]$$

In Figure 10.7, the covariance matrix equation, obtained in equation [10.8], is evaluated for the given values of $\omega_1, \omega_2, D, \mu, \varrho, \alpha$ and the displacement ratios of x, y, x_1, x_2, which are represented by red, green, blue and cyan. The calculated rms ratios are plotted against the speed frequency $v\Omega$ by applying three different speed scales, as follows:

For $0 \leq \lambda \leq 1$: $(\mathrm{v}\Omega)_I = \lambda\omega_1$,
For $1 \leq \lambda \leq 2$: $(\mathrm{v}\Omega)_{II} = (2-\lambda)\omega_1 + (\lambda - 1)\omega_2$,.
For $2 \leq \lambda \leq 3$: $(\mathrm{v}\Omega)_{III} = \omega_2/(3-\lambda)$.

In Figure 10.7, the colored rms ratios are indicated by the black center lines when the special wheelbase $\alpha = 10$ is applied. The colored rms ratios denote $\alpha = \infty$ when they are not indicated by additional black center lines. The dashed lines mean that $\alpha = 0$ is applied. Note that the dashed green line vanishes since there is no angle excitation when the front and rear wheels are close together. Obviously, there are two resonant speeds: the first one around ω_1, where the red and green rms ratios of the vertical and rotational car vibrations become maximal, and the second one near ω_2, where the blue rms ratios of the vertical wheel vibrations are resonant (see Wedig 2016a). The four natural frequencies of the whole vehicle system are shown in Figure 10.7 by $\omega_M = 0.99/s$ and $\omega_J = 1.98/s$ for the translation and rotation of the car body and by $\omega_n = 10.099/s$ and $\omega_m = 10.103/s$ for the vertical vibrations of the front and rear wheels. Similarly, as already shown in Figure 10.4, there are also countably many resonance and absorption speeds. However, in Figure 10.7, these extreme rms ratios are mainly restricted to the speed range of the first resonance, which almost completely vanish in the resonance range of the wheels. This is physically plausible since the car body is spatially extended while both wheels are modeled by mass points without any spatial expansion.

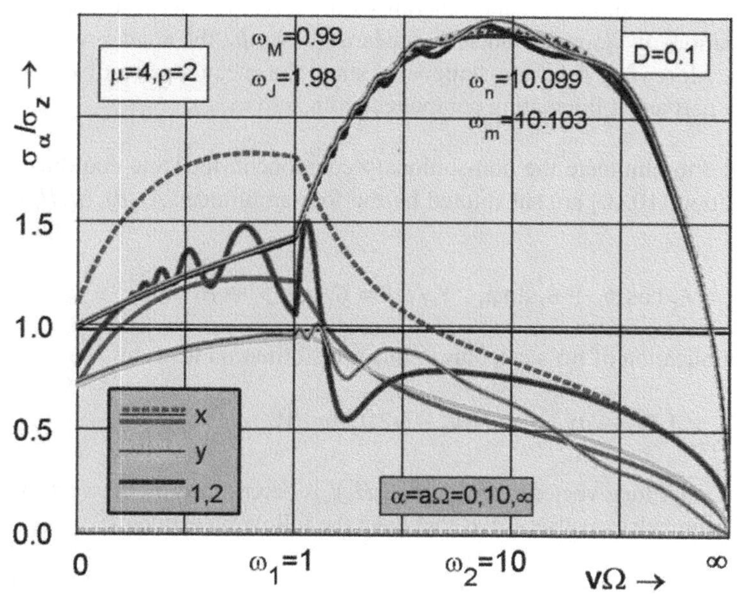

Figure 10.7. *Related rms ratios of all displacements versus the speed frequency for* $\boldsymbol{D = 0.1}$ *and the wheelbases* $\boldsymbol{\alpha = 0, 10}$ *and* ∞*. For a color version of this figure, see www.iste.co.uk/challamel/mechanics3.zip*

10.4. Non-stationary road excitations applying sinusoidal models

The sinusoidal road model shown in Figure 10.1 is generalized by its stochastic version

$$Z_t = \cos\phi_t,\ d\phi_t = \Omega v dt + \sigma dW_t \tag{10.9}$$

where the frequency speed Ωv is additively perturbed by the white noise of intensity σ. This model takes speed fluctuations into account, which are affected by the driving motor and air resistance. The application of Itô's calculus to equation [10.9] gives the slope as

$$dZ_t = -(\Omega v dt + \sigma dW_t)\sin\phi_t - (\sigma^2/2)\cos\phi_t dt$$

This road process is applied to the quarter-car model in a slightly modified form

$$\ddot{Y}_t + 2D\omega_1\dot{Y}_t + \omega_1^2 Y_t = \mu\left(2D\omega_1\dot{Z}_t + \omega_1^2 Z_t\right) \tag{10.10}$$

where μ is the amplitude of the random road excitation. Because of the nonlinear transformation of Gaussian noise by means of $\cos\phi_t$, the road process Z_t is not Gaussian. Moreover, it is non-stationary, since the mean square of $\cos\Omega vt$ has a stationary part and a fluctuating component with zero mean.

In order to eliminate the non-stationary component, the state coordinates Y_t and $\dot{Y}_t$ in equation [10.10] are substituted by the four amplitudes A_t, B_t, G_t, H_t by means of

$$Y_t = A_t\cos\phi_t + B_t\sin\phi_t,\ \dot{Y}_t/\omega_1 = G_t\cos\phi_t + H_t\sin\phi_t \tag{10.11}$$

The application of Itô's calculus to this transformation leads to

$$d\vec{V}_t = \left(A\vec{V}_t - \mu\vec{x}\right)dt + \left(B\vec{V}_t - \mu 2D\vec{y}\right)\sigma\, dW_t \tag{10.12}$$

where the amplitude vector $\vec{V}_t = (A_t\ B_t G_t H_t)'$ is determined by the vectors $\vec{x}$, $\vec{y}$ and the skew-symmetric matrices A, B that have the following elements:

$$A = \begin{bmatrix} \sigma^2/2 & -\Omega v & \omega_1 & 0 \\ \Omega v & \sigma^2/2 & 0 & \omega_1 \\ -\omega_1 & 0 & \sigma^2/2 - 2D\omega_1 & -\Omega v \\ 0 & -\omega_1 & \Omega v & \sigma^2/2 - 2D\omega_1 \end{bmatrix},\quad \vec{V}_t = \begin{bmatrix} A_t \\ B_t \\ G_t \\ H_t \end{bmatrix}$$

$$B = \begin{bmatrix} 0 & -1 & 0 & 0 \\ 1 & 0 & 0 & 0 \\ 0 & 0 & 0 & -1 \\ 0 & 0 & 1 & 0 \end{bmatrix}, \qquad \vec{x} = \begin{bmatrix} 0 \\ 0 \\ \omega_1 - D\sigma^2 \\ 2D\Omega\text{v} \end{bmatrix}, \qquad \vec{y} = \begin{bmatrix} 0 \\ 0 \\ 0 \\ 1 \end{bmatrix}$$

Numerical integration of the transformed vector equation [10.12] is performed by means of the Euler scheme with the time step $\omega_1 \Delta t = 10^{-3}$ for $N =\cdot 10^7$ samples (see Maruyama 1955). According to Kloeden (1995), the Wiener increment in equation [10.12] is approximated by $\Delta W_n = N_n \sqrt{\Delta t}$, where the numbers N_n are normally distributed with zero mean $E(N_n) = 0$ and unit square mean $E(N_n^2) = 1$.

The parameter values selected are $D = 0.1$, $\sigma/\sqrt{\omega_1} = 0.2$ and $\mu = 1$. In Figure 10.8, the simulation results of the displacement means $E(A_t)$ and $E(B_t)$, as well as the density distributions $p(a)$ and $p(b)$ are plotted against the related speed frequency $\nu = \Omega\text{v}/\omega_1$. The $p(a)$ density is indicated by blue and the $p(b)$ density by green. Both densities show the resonance effect when the vehicle is driven at the resonant speed $\nu = 1$. For further growing speed, the densities of both displacements are concentrated around the zero axis, known as self-centering in rotor dynamics.

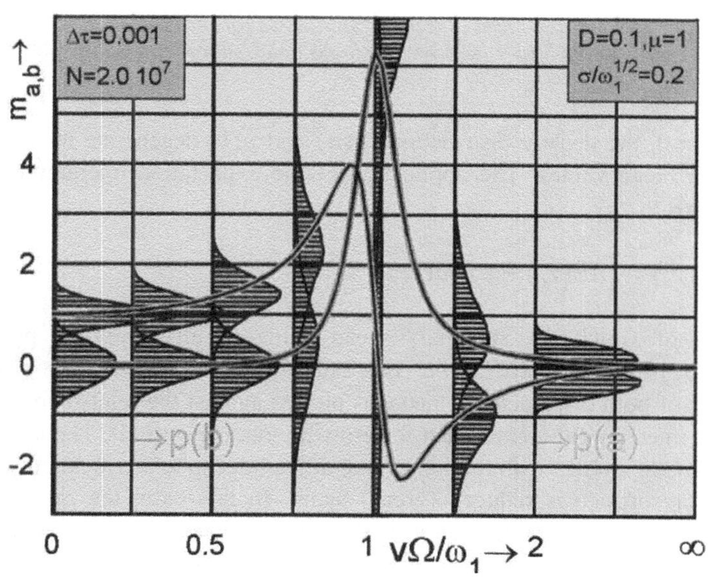

Figure 10.8. *Mean amplitudes and densities of the displacements A_t and B_t versus frequency speed related to the natural frequency of the car. For a color version of this figure, see www.iste.co.uk/challamel/mechanics3.zip*

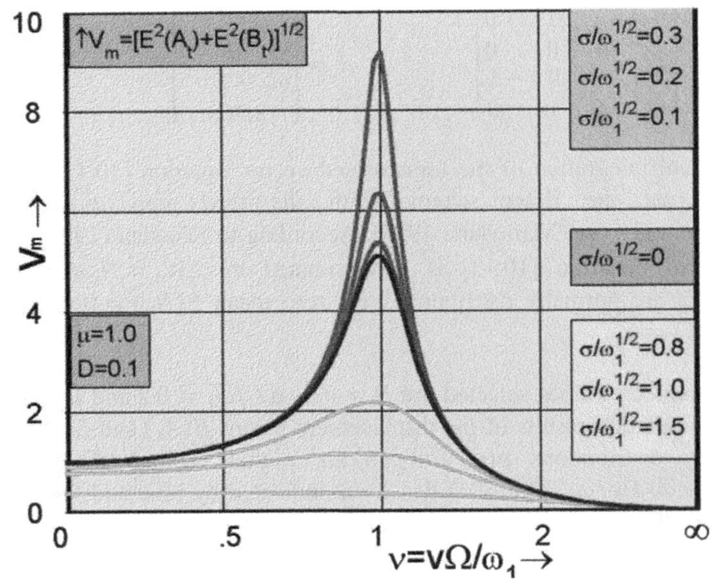

Figure 10.9. *Resonance of the resultant mean amplitude for growing noise that increases (red) first and then decreases (green) again. For a color version of this figure, see www.iste.co.uk/challamel/mechanics3.zip*

For $\sigma = 0$, the density distributions $p(a)$ and $p(b)$ degenerate to delta needles around the mean values. The application of the expectation operator to equation [10.12] leads to

$$dE(\vec{V}_t) = \left[AE(\vec{V}_t) - \mu\vec{x}\,\right]\mathrm{dt} \qquad [10.13]$$

by means of which the stationary mean value of all amplitude processes is calculated. In Figure 10.9, the resultant mean amplitude $V_m = [E^2(A_t) + E^2(B_t)]^{1/2}$ of both displacement means is plotted against the related speed, first for zero noise intensity (blue) and then for growing intensities (red). The latter leads to stable mean amplitudes with resonance magnifications up to a critical noise intensity where the resonance is reduced (green) again. In this case, the mean amplitude becomes unstable. The stability in mean can be investigated using the eigenvalues of the matrix A, denoted immediately after equation [10.12]. In section 10.5, the almost sure stability of the transformed system [10.12] is investigated using Lyapunov exponents.

10.5. Resonance reduction or induction by means of colored noise

The resonance behavior of the amplitude processes is changed when the velocity perturbation by white noise $\sigma \dot{W}_t$ in equations [10.9] and [10.12] is replaced by the more realistic perturbation by means of filtered or colored noise Ψ_t as follows:

$$\dot{Y}_t = \omega_1 X_t, \qquad \dot{\Phi}_t = \Omega \text{v} + \omega_g \Psi_t, \quad d\Psi_t = -\omega_g \Psi_t dt + \sigma d \qquad [10.14]$$

$$\dot{X}_t = -\omega_1(2DX_t + Y_t) + \mu\left[\omega_1 \cos\Phi_t - 2D\left(\Omega \text{v} + \omega_g \Psi_t\right)\sin\Phi_t\right] \qquad [10.15]$$

The stationary filtered noise Ψ_t in equation [10.14] is normally distributed with zero mean $E(\Psi_t) = 0$ and mean square $E(\Psi_t^2) = \sigma_\psi^2 = \sigma^2/\left(2\omega_g\right)$. The application of the amplitude vector $\vec{V}_t$ to the system (Y_t, X_t) leads to the transformed equation

$$d\vec{V}_t = (C + \omega_g B \Psi_t)\vec{V}_t dt + \mu\left(\vec{z} - 2D\omega_g \Psi_t \vec{y}\right)dt \qquad [10.16]$$

where $\vec{V}_t$, B and $\vec{y}$ remain unchanged, while C and $\vec{z}$ are determined by

$$C = \begin{bmatrix} 0 & -\Omega \text{v} & \omega_1 & 0 \\ \Omega \text{v} & 0 & 0 & \omega_1 \\ -\omega_1 & 0 & -2D\omega_1 & -\Omega \text{v} \\ 0 & -\omega_1 & \Omega \text{v} & -2D\omega_1 \end{bmatrix}, \qquad \vec{z} = \begin{bmatrix} 0 \\ 0 \\ \omega_1 \\ -2D\Omega \text{v} \end{bmatrix}.$$

Note that, in equation [10.16], the term $\Psi_t \vec{V}_t$ is nonlinear with the increment

$$d(\Psi_t \vec{V}_t) = \left[(C - \omega_g I)\Psi_t \vec{V}_t + \omega_g B \Psi_t^2 \vec{V}_t + \mu(\Psi_t \vec{x} - 2D\omega_g \Psi_t^2 \vec{y})\right]dt + \vec{V}_t \sigma dW_t,$$

which couples the system vector $\vec{V}_t$ to the higher potencies of the perturbation Ψ_t. This coupling is approximately removed by means of the Gaussian closure $E(\Psi_t^2 \vec{V}_t) = E(\Psi_t^2)E\left(\vec{V}_t\right)$, which leads to the stationary moments equations (see Lin and Cai 2003):

$$CE\left(\vec{V}_t\right) + \omega_g BE\left(\Psi_t \vec{V}_t\right) = -\mu \vec{z}, \qquad E(\Psi_t^2 \vec{V}_t) = E(\Psi_t^2)E\left(\vec{V}_t\right),$$

$$\omega_g \sigma_\psi^2 BE\left(\vec{V}_t\right) + \left(C - \omega_g I\right)E\left(\Psi_t \vec{V}_t\right) = \mu 2D\omega_g \sigma_\psi^2 \, \vec{y}, \qquad E(\Psi_t^2) = \sigma_\psi^2,$$

where I is the unit matrix and $\sigma_\psi^2 = E(\Psi_t^2)$ denotes the mean square of the frequency speed perturbation by filtered noise.

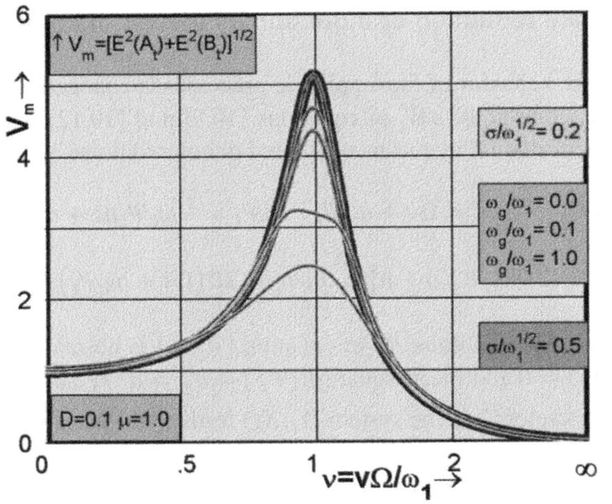

Figure 10.10. *Resonance reduction of the mean amplitude $\boldsymbol{V_m}$ for growing noise intensity (green–blue) and increasing perturbation bandwidth. For a color version of this figure, see www.iste.co.uk/challamel/mechanics3.zip*

Numerical evaluations of the above mean amplitude equations are shown in Figure 10.10, where the resultant mean amplitude V_m is plotted against the related vehicle velocity ν for $D = 0.1$, $\mu = 1$ and the noise intensities $\sigma/\sqrt{\omega_1} = 0.2$ (green) and $\sigma\sqrt{\omega_1} = 0.5$ (blue). For $\omega_g = 0$, the green and blue lines coincide with the deterministic result of vanishing perturbations. For the growing bandwidth of the perturbation, the resonance is reduced. The resonance reduction becomes stronger for increasing noise intensity σ and for increasing low-pass frequency ω_g. This effect can be physically explained by the fact that high frequencies are filtered out and only low frequencies are retained in the perturbation. However, perturbations with low speed frequencies vΩ do not contribute to the resonance effect.

The resonance reduction shown in Figure 10.10 is completely inverted when the high-pass perturbation is applied instead of low-pass noise:

$$Z_t = \cos\phi_t, \qquad \dot{\Phi}_t = \Omega\mathrm{v} + \dot{\Psi}_t, \qquad d\Psi_t = -\omega_g\Psi_t dt + \sigma dW_t \qquad [10.17]$$

where white noise $\sigma\dot{W}_t$ in equation [10.9] is now replaced by $\dot{\Psi}_t$, as defined in equation [10.17]. Consequently, perturbations with low frequencies are filtered out, and high frequencies are retained. The substitution of equation [10.17] into equation [10.10] leads to

$$d\phi_t = (\Omega v - \omega_g\Psi_t)dt + \sigma dW_t,\ d\Psi_t = -\omega_g\Psi_t dt + \sigma dW_t,$$

$$dX_t = -\omega_1(2DX_t + Y_t)dt + \mu\big[(\omega_1 - D\sigma^2)\cos\phi_t + \big(\omega_g\Psi_t - \Omega\text{v}\big)\sin\phi_t\big]dt$$

$$dY_t = \omega_1 X_t dt. -\mu 2D\sigma\sin\phi_t dW_t$$

The application of the amplitudes in equation [10.11] gives the vector equation

$$d\vec{V}_t = \ \big(A - \omega_g B\Psi_t\big)\vec{V}_t\text{dt} + \mu\big(\vec{x} + 2D\omega_g\Psi_t\vec{y}\big)dt + \big(B\vec{V}_t - 2D\mu\vec{y}\ \big)\sigma dW_t,$$

where the vector $\vec{V}_t$ contains the four amplitude processes. The matrices A and B and the vector $\vec{y}$ have already been described in section 10.3.

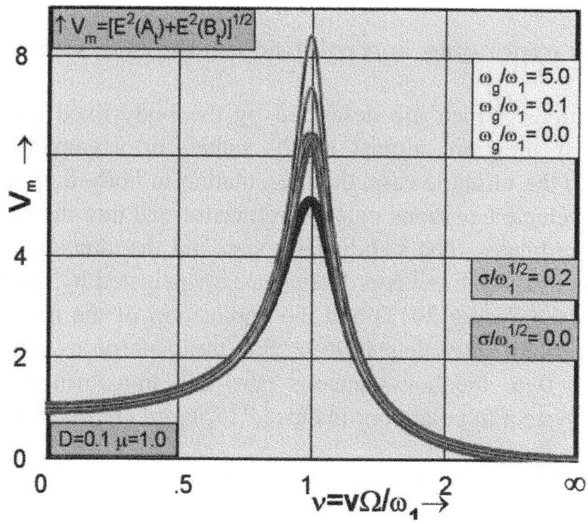

Figure 10.11. *Resonance induction from blue to red resonance peaks for growing noise and to green ones by increasing perturbation bandwidth. For a color version of this figure, see www.iste.co.uk/challamel/mechanics3.zip*

The nonlinear term $\Psi_t\vec{V}_t$ couples the amplitude vector to the higher potencies of the perturbation Ψ_t. This coupling is approximately removed by means of the Gaussian closure $E(\Psi_t^2\vec{V}_t) = E(\Psi_t^2)E\big(\vec{V}_t\big)$, which leads to

$$AE\big(\vec{V}_t\big) - \omega_g BE\big(\Psi_t\vec{V}_t\big) = -\mu\vec{x},\ E(\Psi_t^2\vec{V}_t) = E(\Psi_t^2)E\big(\vec{V}_t\big),$$

$$BE\big(\vec{V}_t\big)\sigma^2/2 + \big(A - \omega_g I\big)E\big(\Psi_t\vec{V}_t\big) = \mu D\vec{y}\sigma^2,\ E(\Psi_t^2) = \sigma_\Psi^2.$$

Numerical evaluations of these moment equations are shown in Figure 10.11, where the mean amplitude V_m is plotted against the related vehicle speed v for $D = 0.1$ and $\mu = 1$. The blue line denotes $\sigma = 0$ of vanishing speed perturbations. For $\sigma/\sqrt{\omega_1} = 0.5$ and $\omega_g/\omega_1 = 0$, we obtain the red line overlaid by the green one, which coincides with the white noise case shown in Figure 10.9. For $\omega_g/\omega_1 = 0.1$ and $\omega_g/\omega_1 = 5$ of increasing bandwidth of the speed perturbation, the resonance peak grows further. This is typical of the parameter resonance when the mean excitation frequency Ωv is around the double eigenfrequency $2\omega_1$ of the quarter-car model. The parameter resonance coincides with the effects of the internal resonance with the frequency ratio of 1:2.

10.6. Lyapunov exponents and rotation numbers in vehicle dynamics

Vehicle vibrations, which are described by the body-fixed coordinates [10.11], physically exist if they are almost surely stable, or asymptotically stable with probability one. In the unstable case, the presentation in body-fixed coordinates is not possible and the vehicle equations must be retransformed into their original equations in space-fixed coordinates. The stability in mean and the almost sure stability of the vehicle vibrations in body-fixed coordinates are investigated by means of projections on hyperspheres (see Wedig 2017) and the application of the multiplicative ergodic theorem introduced by Oseledets (1968). For these purposes, equation [10.12] is reduced with $\mu = 0$ to the homogeneous part, and transformed by means of the two-dimensional system of polar coordinates (P_t, Γ_t) and (Q_t, θ_t) into the equations

$$A_t = P_t \cos \Gamma_t:\ dP_t = [\omega_1 \cos(\Theta_t - \Gamma_t)\, Q_t + \sigma^2 P_t]dt,$$

$$B_t = P_t \sin \Gamma_t:\ d\Gamma_t = [\omega_1 \sin(\Theta_t - \Gamma_t)\ Q_t/P_t + \Omega v]dt + \sigma dW_t$$

$$G_t = Q_t \cos \Theta_t: dQ_t = [-\omega_1 \cos(\Theta_t - \Gamma_t)\, P_t + (\sigma^2 - 2D\omega_1)Q_t]dt,$$

$$H_t = Q_t \sin \Theta_t: d\Theta_t = [\omega_1 \sin(\Theta_t - \Gamma_t)\, P_t/Q_t + \Omega v]dt + \sigma dW_t$$

The equations given above project the displacement amplitudes (A_t, B_t) and the velocities (G_t, H_t) on two circles with radii (P_t, Q_t) and angles (Γ_t, Θ_t). A second application of polar coordinates eliminates the two radii (P_t, Q_t) and projects the entire motion on a hypersphere with one radius R_t and the three angles Ψ_t, Δ_t and Σ_t, where the difference angle Δ_t and the sum angle Σ_t are introduced as follows:

$$P_t = R_t \cos\Psi_t:\ dR_t = [\sigma^2 - D\omega_1(1 - \cos 2\Psi_t)]R_t dt \qquad [10.18]$$

$$Q_t = R_t \sin\Psi_t: d\Psi_t = -\omega_1(D\sin 2\Psi_t + \cos\Delta_t)dt \qquad [10.19]$$

$$\Delta_t = \theta_t - \Gamma_t: d\Delta_t = \omega_1 \sin\Delta_t(\cot\Psi_t - \tan\Psi_t)dt \quad [10.20]$$

$$\Sigma_t = \theta_t + \Gamma_t: \ d\Sigma_t = \omega_1 \sin\Delta_t(\cot\Psi_t + \tan\Psi_t)dt + 2\omega_e dt + 2\sigma dW_t$$

Note that the angle equations [10.19] and [10.20] are decoupled and can be solved separately. Equation [10.18] is integrated and leads to the top Lyapunov exponent

$$\lambda_{top} = \lim_{t\to\infty} \frac{1}{t} \ln(R_t/R_0) = \sigma^2 - D\omega_1 \left(1 - \lim_{t\to\infty} \frac{1}{t} \int_0^t \cos 2\Psi_\tau d\tau\right) \quad [10.21]$$

According to equation [10.21], the top Lyapunov exponent is determined by the noise intensity, the system damping and the time average of the cosine of the stability angle Ψ_t, which is coupled with the difference angle Δ_t. Both the angles are determined by equations [10.19] and [10.20], which are nonlinear, noise-free and decoupled from the sum angle Σ_t, where additive noise is still active.

In Figure 10.12, equations [10.19] and [10.20] are numerically evaluated for the under- and overcritical damping $D = 0.5$ and $D = 1.5$ on the left half and the right half, respectively. The angle solutions are 2π –periodic in Δ and π –periodic with respect to Ψ. For $D \leq 1$, Figure 10.12 shows the stationary limit cycles of both angles $(\Lambda_\tau, \Psi_\tau)$. They can be calculated without any transient time behavior for arbitrary initial values by applying an Euler scheme, for example with the time step $\Delta\tau = \omega_1 \Delta t = 0.001$.

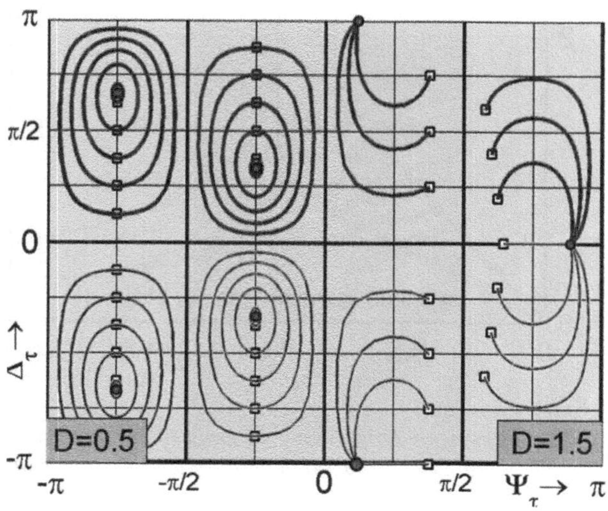

Figure 10.12. *Limit cycles and transient projections of angles Δ_τ, Ψ_t and their ends around red centers and in fix-points, respectively for $D > 1$ and $D < 1$. For a color version of this figure, see www.iste.co.uk/challamel/mechanics3.zip*

All the limit cycles shown in Figure 10.12 are symmetric with respect to $\Psi = \pm\pi/4$. Because of this symmetry, the time average in equation [10.21] can also be obtained by four singular limit cycles around $\Psi = \pm\pi/4$ and $\cos\Delta = \pm D$, which represent the center points indicated in red. The substitution of these center values into equation [10.21] leads to one top Lyapunov exponent and two rotation numbers as follows:

$$\lambda_{top} = \sigma^2 - D\omega_1 \text{ and } \rho_{\Gamma,\theta} = \Omega \text{v} \pm \omega_1\sqrt{1-D^2} \qquad \text{for } D \le 1 \qquad [10.22]$$

The rotation numbers are time averages of rotating angles applied to the angle increments $d\Gamma_t$ and $d\Theta_t$. They are calculated by the values $\sin^2\Delta = 1 - D^2$ and $Q/P = \tan\Psi = \pm 1$. Subsequently, equations [10.19] and [10.20] are evaluated for the overcritical damping value $D = 1.5$, where transient solutions are obtained instead of limit cycles. Starting with any initial values, both angles $(\Lambda_\tau, \Psi_\tau)$ move to the stationary fix-points $\Delta = 0, \pm\pi$ and $\sin 2\Psi = \pm 1/D$. The substitution of these stationary angle values leads to two Lyapunov exponents and the rotation number:

$$\lambda_{1,2} = \sigma^2 - \omega_1\left(D \mp \sqrt{D^2-1}\right) \text{ and } \rho_{\Gamma,\theta} = \Omega \text{v} \qquad \text{for } D \ge 1 \qquad [10.23]$$

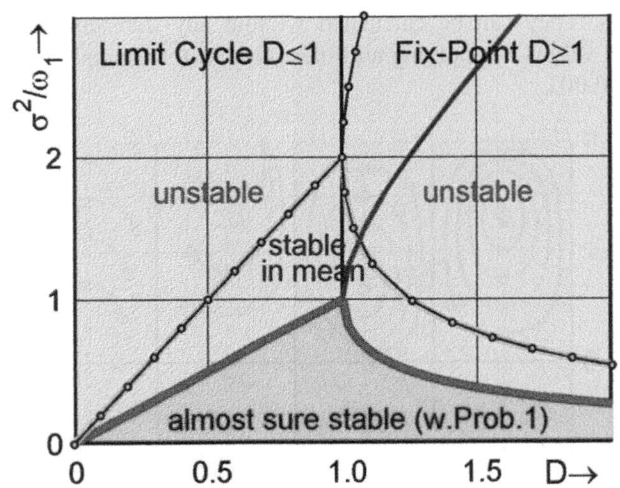

Figure 10.13. *Boundaries for almost sure stability and stability in mean. Critical intensity of noise perturbation plotted against system damping. For a color version of this figure, see www.iste.co.uk/challamel/mechanics3.zip*

In Figure 10.13, the stability boundaries are obtained by plotting the critical noise intensity σ_{cr}^2/ω_1 against vehicle damping. The stability limits are

$$\sigma_{cr}^2/\omega_1 = D, \qquad \text{for} \quad D \leq 1,$$

$$\sigma_{cr}^2/\omega_1 = D - \sqrt{D^2 - 1}, \qquad \text{for} \quad D \geq 1$$

which are obtained for $\lambda_{top} = 0$. In Figure 10.13, the stability region is indicated in green. Inside the green region, the stationary solution of equation [10.12] is asymptotically stable with probability one, or almost surely stable. Above the green range, the vector solution is unstable. Obviously, we need linearly increasing damping in order to stabilize the system for growing noise. However, this effect only holds up to $D = 1$. Overcritical damping is less effective. For $D \rightarrow \infty$, the vehicle and the road are rigidly coupled. Hence, damping and stabilization are no longer possible.

For an extended stability investigation, the above polar coordinate system (P_t, Γ_t) and (Q_t, θ_t) is also applied to the mean value equation [10.13]. For $\mu = 0$, this leads to the same stability equations for the angles (Γ_t, θ_t) and radii (P_t, Q_t), except that σ^2 in the radii equations is replaced by $\sigma^2/2$, since there is no noise in equation [10.13] for $E(dW_t) = 0$. Thus, the radius equation [10.18] is replaced by

$$dR_t = [\sigma^2/2 - D\omega_1(1 - \cos 2\Psi_t)]R_t dt \qquad [10.24]$$

and the Lypaunov exponents of the mean value solutions are calculated to obtain

$$\lambda_{m,} = \sigma^2/2 - D\omega_1, \quad \text{for} \quad D \leq 1$$

$$\lambda_{m,} = \sigma^2/2 - \omega_1\left(D \mp \sqrt{D^2 - 1}\right), \qquad \text{for } D \geq 1$$

The rotation numbers of the mean value solutions remain unchanged. In Figure 10.13, the stability boundaries of the mean value solutions are indicated by the blue lines. Extensions to stability in the p-th mean and relations to almost sure stability, where $p = 0$, are given by Arnold (2003). Applications of the p-th mean calculation to the stochastic oscillator under multiplicative noise are given by Wedig (1988).

10.7. Concluding remarks and main new results

The first part of this chapter investigated multi-body half-car models rolling on random road profiles, which are described by stationary Gaussian processes, with time delays affected by the center distance of the front and rear wheels. New results presented in Figure 10.5 showed many resonance and absorption speeds plotted against all possible wheelbases of the half-car model. The resonance and absorption speeds are calculated for the limiting case in which damping tends to zero, similar to the eigenfrequency calculation in vibration systems.

In the second part of this chapter, the stationary Gaussian ground excitation is replaced by a noisy sinusoidal model, where the time frequency given by the car velocity times the road frequency is additively perturbed by white or colored noise. This is the so-called Dimentberg Wedig model (see Lin and Cai 2004). It is non-Gaussian but more realistic since the road level is strictly bounded by the sinusoidal model. However, its time derivative is unlimited as derived by applying Itô's calculus to the increment [10.9], in order to obtain the slope process of the road. The non-stationarity is eliminated by introducing stationary amplitude processes into the system.

New results of the quarter-car model were presented in Figure 10.11, where the resultant mean value of two amplitude deflections was plotted against the speed frequency of the car. The amplitude frequency curve obtained remains almost unchanged outside the resonance. However, the resonance amplification is considerably increased by applying white noise perturbations. The resonance induction is even stronger when we apply perturbations generated by high-pass filters, which eliminate low frequencies with increasing filter frequency, instead of white noise. This effect corresponds to the parameter resonance or the internal resonance of 1:2.

The introduction of stationary amplitude processes leads to transformed equations of motion perturbed by white noise in additive and multiplicative form. The almost sure stability (stability with probability one) of the transformed equations is investigated by means of the multiplicative ergodic theorem. The application of the expectation operator to the transformed equations leads to the linear mean value equations with constant coefficients. Their eigenvalues determine the corresponding stability in mean. This chapter investigated the stability in mean by applying the projection on hyperspheres, in order to obtain the moment Lyapunov exponents.

10.8. References

Arnold, L. (1974). *Stochastic Differential Equations*. John Wiley & Sons, New York.

Arnold, L. (2003). *Random Dynamical Systems*. Springer Verlag, New York.

Davis, B.R. and Thompson, A.G. (2001). Power spectral density of road profiles. *Vehicle System Dynamics*, 35(6), 409–441.

Den Hartog, J.P. (1934). *Mechanical Vibrations*. McGraw-Hill Book Company, New York.

Di Poala, M. and Pirotta, A. (2001). Time delay induced effects on control of linear systems under random excitation. *Probabilistic Engineering Mechanics*, 16(1), 48–52.

Doods, C.J. and Robson, D.J. (1973). The description of road surface roughness. *Journal Sound and Vibration*, 31, 175–183.

Elishakoff, I. (1983). *Probabilistic Methods in the Theory of Structures*. John Wiley & Sons, New York.

Itô, K. (1951). On stochastic differential equations. *American Mathematical Society*, 4, 1–51.

Kloeden, P. and Platen, P.E. (1995). *Numerical Solution of Stochastic Differential Equations: A Review*. Springer, Heidelberg.

Klotter, K. (1955). *Schwingungen, Hütte, Theoretische Grundlagen*, 4(28). Berlin, Verlag Ernst & Sohn, 580–582.

Klotter, K. (1978). *Technische Schwingungslehre*, Bd.1&2. Springer-Verlag, Heidelberg.

Lin, Y.K. and Cai, G.Q. (2004). *Probablistic Structural Dynamics*. McGraw-Hill, New York.

Maruyama, G. (1955). Continuous Markov process and stochastic equations. *Rendiconti del Circolo Matematico di Palermo*, 4, 48–90.

Oseledets, V.I. (1968). A multiplicative ergodic theorem: Lyapunov characteristic numbers for dynamical systems. *Transactions of the Moscow Mathematical Society*, 19, 197–231.

Popp, K. and Schiehlen, W.O. (1993). *Fahrdynamik*. Teubner, Stuttgart.

Sobczyk, K., MacVean, D.B., Robson, J.D. (1977). Response to profile-imposed excitation with randomly varying transient velocity. *Journal of Sound and Vibration*, 52(1), 37–49.

Wedig, W.V. (1988). Lyapunov exponents of stochastic systems and related bifurcation problems. In *Stochastic Structural Dynamics – Progress in Theory and Applications*, Ariaratnam S.T., Schueller G.I. and Elishakoff I. (eds). Elsevier Applied Science, London.

Wedig, W.V. (2003). Dynamics of cars driving on stochastic roads. In *Proc. of Computational Stochastic Mechanics*, Spanos, P.D. and Deodatis, G. (eds). Millpress, Rotterdam.

Wedig, W.V. (2012). Digital simulation of road-vehicle systems. *Probabilistic Engineering Mechanics*, 27, 82–87.

Wedig, W.V. (2016a). New resonances and velocity jumps in non-linear road vehicle dynamics. *IUTAM Symposium Analytical Methods in Nonlinear Dynamics*, Procedia IUTAM, 19, 209–219.

Wedig, W.V. (2016b). Jump phenomena in road-vehicle dynamics. *Int. Journal of Dynamics and Control*, 4, 213–220.

Wedig, W.V. (2017). Lyapunov exponents and rotation numbers in rotor and vehicle dynamics. *Procedia Engineering*, 199, 875–881.

Wedig, W.V. (2020). Turbulent travel speeds in nonlinear vehicle dynamics. *Journal Nonlinear Dynamics*, 100(1), 147–158.

11

Order Statistics Approach to Structural Optimization Considering Robustness and Confidence of Responses

A method for the worst-case analysis of structures is presented based on a random sampling approach and order statistics. The approximate worst value of the response is obtained with pre-assigned accuracy using the distribution-free tolerance interval of order statistics. A multi-objective optimization formulation is presented for robust design relating the order of response to the robustness level. The methods are applied to robust optimum design of building frames subjected to seismic motions.

11.1. Introduction

Among many methods for considering the uncertainty of parameters in the practical design process, reliability-based design (RBD) is a standard approach that can incorporate various types of uncertainty, including those in material, geometry and external loads. However, RBD usually involves assumptions on probability density functions, and a small probability of failure is sensitive to the parameter variation due to the behavior of the tail of the distribution, especially in the case of unbounded support. The deterministic worst-case (WC) approach with bounded support is also widely used, because it is simpler, less sensitive, and yields the same results as a probabilistic methodology under appropriate assumptions (Elishakoff and Ohsaki 2010).

Chapter written by Makoto YAMAKAWA and Makoto OHSAKI.

In some special cases, we can find the exact or moderately accurate worst value of structural responses (Ben-Tal *et al.* 2009; Elishakoff and Ohsaki 2010); however, in many practical problems, finding such worst value is impossible or requires large computational cost even when simple interval variables are used for representing uncertainty (Ben-Tal *et al.* 2009; Ohsaki and Katsura 2012). Note that the exact worst value may correspond to an extreme rare event that has probability 0 from the viewpoint of probability theory. For example, it is impossible and/or unreasonable to ensure the safety of a building structure against any unexpected natural disasters. For some cases, it may be important to minimize the median and/or percentile values of the response. Hence, we can justify relaxing from the worst value to a worse value.

The key concept of the method presented in this chapter is the estimation of robustness by order statistics. Robustness is defined by non-parametric tolerance intervals for various percentile and confidence levels (Yamakawa and Ohsaki 2013; Ohsaki *et al.* 2019). Structural responses are regarded as random variables; however, we only assume the existence of their probability density function, and do not know the type of distribution of the structural response, which is the case for most probabilistic approaches such as the Monte Carlo method. Furthermore, we need not make more specific assumptions such as the normal distribution, time-wise stationarity, space-wise homogeneity, ergodicity of the process and the Markov property. The method can be placed between a pure probabilistic methodology and a fully deterministic technique, and it is easily implemented using a deterministic numerical code or existing simulation software. The validity of the method is demonstrated using the examples of building frames, and it is shown that a relatively small set of samples is enough to make a prediction, even for noisy and non-smooth structural responses under uncertain parameters.

11.2. Overview of order statistics

Some recent studies linked a robust design optimization or a reliability-based design optimization with non-parametric order statistics (Ohsaki *et al.* 2019; Jekel and Haftka 2020), which is briefly introduced in this section.

11.2.1. *Definition of order statistics*

Let $\boldsymbol{\theta} = (\theta_1, \theta_2, \ldots, \theta_t) \in \Omega$ denote a t-dimensional random variable vector, which corresponds to uncertain parameters such as yield stress, Young's modulus, cross-sectional area and external load, where $\Omega \subset \mathbb{R}^t$ is the prescribed continuous and bounded set. The representative response of a structure is denoted by $g(\boldsymbol{\theta})$. Let

$\boldsymbol{\theta}_1, \boldsymbol{\theta}_2, \ldots, \boldsymbol{\theta}_n$ denote a set of n random variable vectors that are assumed to be independent and generated using the same probability distribution on Ω. Accordingly, the representative responses denoted by $Y_1 = g(\boldsymbol{\theta}_1)$, $Y_2 = g(\boldsymbol{\theta}_2)$, ..., $Y_n = g(\boldsymbol{\theta}_n)$ are also assumed to be independent and identically distributed random variables (iidrvs), for which the cumulative distribution function (cdf) is assumed to be continuous and denoted by

$$F_Y(y) = \Pr\{Y \leq y\} = \Pr\{\boldsymbol{\theta} \in \Omega \mid g(\boldsymbol{\theta}) \leq y\}. \qquad [11.1]$$

These random variables $Y_1, Y_2, \cdots, Y_n$ are renumbered in decreasing order as $Y_{1,n} \geq Y_{2,n} \geq \ldots \geq Y_{n,n}$, where the kth largest response $Y_{k,n}$ among n samples is referred to as the kth order statistic (David and Nagaraja 2003). Note that they are renumbered in increasing order in usual order statistics. We align them in decreasing order because we are interested in large responses.

11.2.2. *Tolerance intervals and confidence intervals of quantiles*

A content ratio by order statistic $F_Y(Y_{k,n})$ is also regarded as a random variable. It is known that $F_Y(Y_{k,n})$ is equal in distribution to the kth order statistic for the n samples from a standard uniform distribution $U(0,1)$. The probability density function (pdf) of $F_Y(Y_{k,n})$ is given by

$$f_{k,n}(t) = \frac{t^{n-k} t^{k-1}}{B(n-k+1,k)} \qquad [11.2a]$$

$$B(a,b) = \int_0^1 t^{a-1}(1-t)^{b-1} dt, \qquad [11.2b]$$

where $B(\cdot,\cdot)$ denotes the beta function (see, for example, David and Nagaraja 2003). The expectation and variance of $F_Y(Y_{k,n})$ are, respectively, given by

$$\mathbb{E}[F_Y(Y_{k,n})] = \frac{n-k+1}{n+1} \qquad [11.3a]$$

$$\mathrm{Var}[F_Y(Y_{k,n})] = \frac{k(n-k+1)}{(n+1)^2(n+2)}. \qquad [11.3b]$$

Furthermore, for a given real number $\gamma \in (0,1)$, the following equation is derived:

$$\Pr\{F_Y(Y_{k,n}) \geq \gamma\} = \int_\gamma^1 f_{k,n}(t)dt = 1 - I_\gamma(n-k+1,k), \qquad [11.4]$$

where $I_\gamma(\cdot,\cdot)$ is the incomplete beta function that is defined by

$$I_\gamma(a,b) = \int_0^\gamma t^{a-1}(1-t)^{b-1}dt \Big/ \int_0^1 t^{a-1}(1-t)^{b-1}dt. \qquad [11.5]$$

If we select k and n $(1 \leq k \leq n)$ that satisfy equation [11.4] for given α and γ $(0<\alpha<1,\ 0<\gamma<1)$ as

$$\Pr\{F_Y(Y_{k,n}) \geq \gamma\} \geq \alpha \Leftrightarrow I_\gamma(n-k+1,k) \leq 1-\alpha, \qquad [11.6]$$

we state that "the probability of $g(\mathbf{\theta})$ so that its content ratio of γ is less than $Y_{k,n}$ is with α confidence." Note that the right-hand side of equation [11.6] does not depend on the type of $F_Y(\cdot)$, which means that $F_Y(Y_{k,n})$ is a distribution-free variate. If F_Y is a strictly increasing function, the equation $F_Y(y)=\gamma$ $(0<\gamma<1)$ has a unique solution denoted by $q_\gamma = F_Y^{-1}(\gamma)$, which is called the (population) γ-quantile. For example, $q_{1/2}$ is the median of the distribution. Even though F_Y does not strictly increase, we can define the γ-quantile as $q_\gamma = \inf\{y \mid F_Y(y) \geq \gamma\}$, and clearly, $F_Y(q_\gamma)=\gamma$ holds. Thus, equation [11.6] is rewritten as

$$\Pr\{q_\gamma \leq Y_{k,n}\} \geq \alpha \Leftrightarrow I_\gamma(n-k+1,k) \leq 1-\alpha. \qquad [11.7]$$

As stated above, equation [11.7] holds independently of the distribution F_Y. Although the value of the γ-quantile is unknown without the information of distribution, we can observe that the sample value of $Y_{k,n}$ corresponding to the upper bound of the γ-quantile with confidence is $100\alpha\%$. This formulation is known as the one-sided non-parametric tolerance interval (Krishnamoorthy and Mathew 2009). Some results of solving equation [11.6] are summarized in Tables 11.1 and 11.2.

k	1	2	3	4	5	6	7	8	9	10
n	22	38	52	65	78	91	104	116	128	140

Table 11.1. *Sample size requirements for $\alpha = \gamma = 0.9$*

k	1	2	3	4	5	6	7	8	9	10
γ	1.000	1.000	1.000	1.000	0.999	0.998	0.994	0.986	0.969	0.940
k	11	12	13	14	15	16	17	18	19	20
γ	0.894	0.829	0.746	0.647	0.540	0.432	0.331	0.242	0.169	0.113

Table 11.2. *Content ratio for order statistics of $\alpha = 0.9$, $n = 150$*

11.3. Robust design

11.3.1. *Overview of the robust design problem*

The upper bound of the γ-quantile of response for specified confidence can be quantified using the theory of order statistics. A structure has a high robustness if it is designed to satisfy a small upper bound for the kth worst representative response that obviously increases as γ is increased. Hence, higher γ and/or lower q_γ may be regarded as representing the robustness level. Figure 11.1 shows the cdfs of a representative response corresponding to some designs. In Figure 11.1(a), design (A) has a smaller γ-quantile for any value of γ and has a higher γ for any γ-quantile value than design (B), which means that the kth worst response of design (A) is likely to be smaller than that of design (B); thus, design (A) is superior to design (B) in regards to robustness. Unfortunately, this clear superiority does not always hold, and generally depends on the robustness level. In Figure 11.1(b), design (C) has a lower γ-quantile to γ_1, and a higher γ-quantile to γ_2 than design (D); therefore, design (C) is said to be non-inferior to design (D). This kind of situation often occurs; for example, a design that minimizes the worst value may not be the same as the design that minimizes the median value. Anyway, we can regard the γ-quantile q_γ as an index of robustness level; hence, the kth order statistic $Y_{k,n}$ can also be linked to the robustness level through equation [11.7].

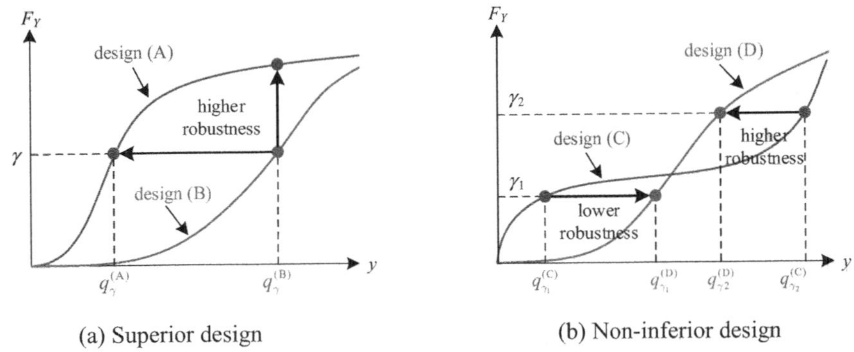

Figure 11.1. *Cumulative distribution function and robustness level. For a color version of this figure, see www.iste.co.uk/challamel/mechanics3.zip*

11.3.2. *Worst-case-based method*

We first define a robust design problem as a two-level problem, where uncertainty is included in the lower-level optimization problem. Consider a problem of optimizing some structural parameters, for example the cross-sectional geometry, the material parameters and so on. The design variable vector is denoted by $\mathbf{x} = (x_1, x_2, \cdots, x_d)$, where d is the number of design variables. The vector consisting of uncertain parameters is denoted by $\boldsymbol{\theta} = (\theta_1, \theta_2, \ldots, \theta_t) \in \Omega$. For simplicity, uncertainty is only incorporated into the objective function $g = g(\mathbf{x}, \boldsymbol{\theta})$ representing a structural response. Based on the worst-case analysis [11.8], the lower-level optimization problem is formulated for finding the worst value of the structural response as follows:

$$\text{Find} \qquad g_{\max}(\mathbf{x}) = \max_{\boldsymbol{\theta} \in \Omega} g(\mathbf{x}, \boldsymbol{\theta}). \qquad [11.8]$$

The constraints without uncertainty are denoted by $h_i = h_i(\mathbf{x})$ $(i = 1, \cdots, l)$, where l is the number of constraints. Thus, a worst-case-based robust optimization [11.9] is formulated with upper bounds $\overline{h}_i$ $(i = 1, \cdots, l)$ as:

$$\left.\begin{array}{lll} \underset{\mathbf{x}}{\text{Minimize}} & g_{\max}(\mathbf{x}) & \\ \text{subject to} & h_i(\mathbf{x}) \le \overline{h}_i & (i = 1, \cdots, l) \end{array}\right\} \qquad [11.9]$$

11.3.3. *Order statistics-based method*

As stated in the introduction, there are difficulties in solving [11.8], and therefore a random sampling approach can be successfully applied to obtain approximate worst solutions using the order of the objective value (Ohsaki and Katsura 2012), i.e. the *k*th worst value of the samples needs to be minimized. In the algorithm of general random search (Zhigljavsky and Žilinskas 2008), a sequence of random points $\boldsymbol{\theta}_1, \boldsymbol{\theta}_2, \cdots, \boldsymbol{\theta}_n$ on Ω is generated so that the point $\boldsymbol{\theta}_j$ $(1 \leq j \leq n)$ has a prescribed probability distribution P_j. When all the distributions are the same, i.e. $P_j = P$ for all j, and the random variables Θ_j are identically distributed, the algorithm is often called pure random search (PRS). As a result of the application of PRS, we obtain independent n samples $\{\boldsymbol{\theta}_1, \cdots, \boldsymbol{\theta}_n\}$ from a distribution P on Ω and the samples of function values $\{Y_1 = g(\mathbf{x}, \boldsymbol{\theta}_1), \cdots, Y_n = g(\mathbf{x}, \boldsymbol{\theta}_n)\}$ for a given $\mathbf{x}$. We arrange the iidrvs in decreasing order of magnitude and observe the *k*th order statistic $Y_{k,n}$. Instead of [11.9], an order statistics-based robust optimization for minimizing the *k*th order statistic can be formulated as:

$$\left.\begin{array}{lll} \underset{\mathbf{x}}{\text{Minimize}} & Y_{k,n} = Y_{k,n}(\mathbf{x}) & \\ \text{subject to} & h_i(\mathbf{x}) \leq \overline{h}_i & (i = 1, \cdots, l) \end{array}\right\} \qquad [11.10]$$

As discussed above, the estimated value of the maximum response becomes more robust if k is smaller. However, larger robustness leads to a larger estimate of the response for a given $\mathbf{x}$, and they are in trade-off relations. Therefore, it is practically important to compare the solutions with various robustness levels to decide the appropriate value of robustness. For this purpose, an order statistics-based multi-objective optimization [11.11], which minimizes a set of order statistics $Y_{\tilde{k}_1,n}(\mathbf{x}), \ldots, Y_{\tilde{k}_m,n}(\mathbf{x})$, is formulated as:

$$\left.\begin{array}{lll} \underset{\mathbf{x}}{\text{Minimize}} & Y_{\tilde{k}_1,n}(\mathbf{x}), \ldots, Y_{\tilde{k}_m,n}(\mathbf{x}) & \\ \text{subject to} & h_i(\mathbf{x}) \leq \overline{h}_i & (i = 1, \cdots, l) \end{array}\right\} \qquad [11.11]$$

11.4. Numerical examples

11.4.1. *Design response spectrum*

Two numerical examples are presented for seismic design of building frames. The design acceleration response spectrum in Japanese building code used in the examples is formulated as

$$S_\mathrm{a}(T, h) = H(h) S_{\mathrm{a}0} \qquad [11.12a]$$

$$S_{a0} = \begin{cases} 0.96 + 9T & (T < 0.16) \\ 2.4 & (0.16 \le T < 0.864) \\ 2.074 / T & (0.864 \le T) \end{cases} \quad [11.12b]$$

where T is the natural period, h is the damping factor of the building frame and $H(h)$ is the damping correction coefficients calculated by $H(h) = 1.5/(1+10h)$. Examples of $S_a(T,h)$ (m/s^2) for $h = 0.02$, 0.05 and 0.10 are plotted in Figure 11.2.

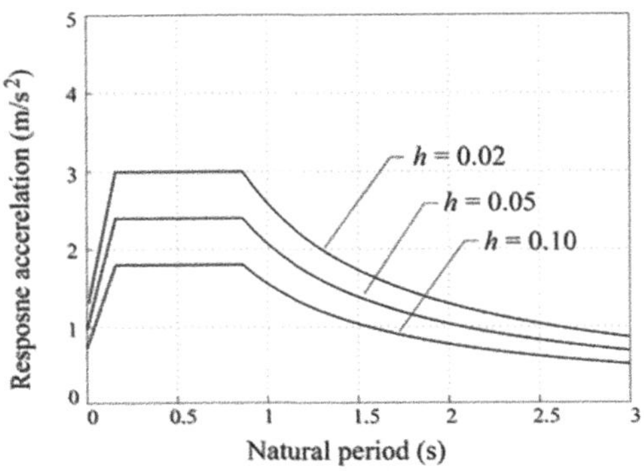

Figure 11.2. *Deign acceleration response spectra for damping factors 0.02, 0.05 and 0.10*

11.4.2. *Optimization of the building frame considering seismic responses*

For earthquake-resistant design of structures, we should appropriately consider the uncertainty in seismic loads to ensure high robustness. For the design response spectra, the acceleration response spectra at the engineering bedrock surface can be specified (Iyengar and Rao 1979); however, there is uncertainty surrounding their phase spectra and ground surface properties. We investigate the applicability of robust optimal design based on order statistics when there are uncertain properties in the phase spectrum, layer thickness and shear wave velocity of the surface ground (Yamakawa *et al.* 2018).

Let us consider a nine-story two-bay rigidly jointed steel frame with pin-jointed buckling-restrained braces, as shown in Figure 11.3(a), on the ground model with layers 1 and 2 from the surface, as shown in Figure 11.3(b). The thicknesses, the shear wave velocity and the mass density of the soil in each layer are denoted by H_m, V_m and ρ_m $(m = 1, 2)$, respectively. The cross-sectional areas of central steel material of nine buckling-restrained braces are taken as the design variables

$x_1, x_2, \ldots, x_9$, for which the upper and lower bounds are $20{,}000$ mm^2 and 0, respectively. The steel materials are assumed to have bilinear stress–strain relations, where Young's modulus is 205 kN/mm^2 and the yield stress is 325 N/mm^2 for beams and columns, and 100 N/mm^2 for braces.

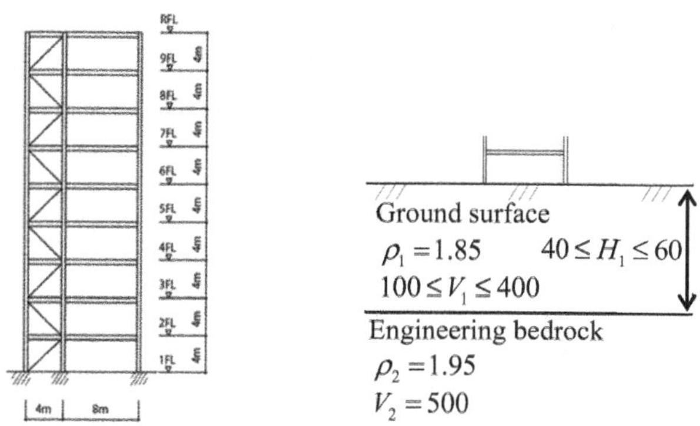

(a) A braced nine-story steel frame (b) Two-layer ground

Figure 11.3. *Building model*

In this example, the phase spectrum of the design seismic motion, layer thickness and shear wave velocity of the surface ground, respectively, have a single uncertain parameter, i.e. the number of uncertain parameters t is three. The phase spectra are given by Hachinohe EW of the Tokachi-oki earthquake in 1968, Tohoku University NS of the Miyagi-oki earthquake in 1978 and Kobe NS of the Hyogoken-nanbu earthquake in 1995, which are indexed by 1, 2 and 3, respectively, and one of them is randomly selected as $\theta_1 = \phi \in \{1,2,3\}$. Dependence of the shear modulus and damping ratios on the strain level is taken into account by an equivalent linear model (Schnabel *et al.* 1972) with the Hardin–Drnevich model (Hardin and Drnevich 2002). The layer thickness and shear wave velocity of the upper surface ground are taken as the uncertain parameters, which are, respectively, denoted by $\theta_2 = H_1$ (m) and $\theta_3 = V_1$ (m/s) as

$$\boldsymbol{\theta} = (\theta_1, \theta_2, \theta_3) = (\phi, H_1, V_1) \in \Omega, \qquad [11.13]$$

$$\Omega = \{(\theta_1, \theta_2, \theta_3) \mid \theta_1 \in \{1,2,3\}, 40 \leq \theta_2 \leq 60, 100 \leq \theta_3 \leq 400\}. \qquad [11.14]$$

The design acceleration response spectrum for 5% damping at the engineering bedrock surface given by equation [11.12] is shown with the bold line in Figure 11.4. The acceleration response spectra considering the uncertain amplification at the ground surface with the Hachinohe EW phase are shown by the dotted lines in Figure 11.4, which shows a strong effect of uncertainty in ground properties on surface ground motions. The response function to be minimized is the maximum inter-story drift angle, denoted by $Y(\mathbf{x}) = g(\mathbf{x}, \theta)$. OpenSees (McKenna 2011) is used for the seismic response analysis of the frame. Each of the beams and columns is modeled by a force-based beam–column element, the section of which is divided into fibers. A truss element is used for the brace. The standard Newmark-β method is used for transient analysis (see Yamakawa *et al.* (2018) for details).

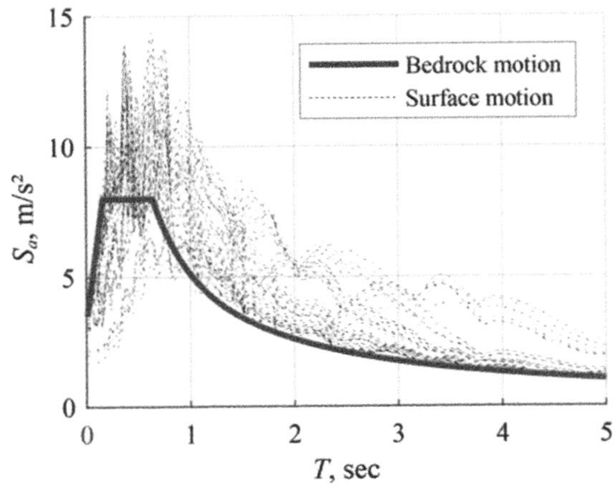

Figure 11.4. *Acceleration response spectra of seismic motions at bedrock and surface with the Hachinohe EW phase. For a color version of this figure, see www.iste.co.uk/challamel/mechanics3.zip*

	x_1	x_2	x_3	x_4	x_5	x_6	x_7	x_8	x_9
ROS	150	198	145	142	108	74	60	200	25
NOS	102	150	133	102	126	94	200	200	115

Table 11.3. *Optimal solutions (cross-sectional areas of braces, unit: 10^2 mm^2)*

	Max (rad)	Mean (rad)	Std. Dev. (rad)	Worst 10% (rad)	$Y_{2,38}$ (rad)	Ratio (%)
ROS	0.023	0.014	0.0025	0.017	0.020	98
NOS	0.030	0.014	0.0042	0.021	0.027	98

Table 11.4. *Summary of observed maximum inter-story drift angles*

We apply the [11.15] to this design problem as follows:

$$\left.\begin{array}{ll}\underset{\mathbf{x}}{\text{Minimize}} & Y_{2,38}(\mathbf{x}) \\ \text{subject to} & 0 \le x_i \le 20000\,\text{mm}^2 \quad (i=1,\cdots,9)\end{array}\right\}$$

The problem is solved by the direct search method (see Yamakawa *et al.* (2015) for details), and the solution is referred to as the robust optimal solution (ROS). The ROS minimizes $Y_{2,38}(\mathbf{x})$ corresponding to $\alpha=\gamma=0.9$ in equation [11.6], as shown in Table 11.1. For comparison, we also find nominal optimal solution (NOS) that minimizes the maximum inter-story drift angle without considering the uncertain amplification of the surface ground, i.e. using seismic motion at the engineering bedrock. Both the ROS and the NOS are listed in Table 11.3. Note that 38 function evaluations of $g(\cdot)$ are needed for one evaluation of $Y_{2,38}(\mathbf{x})$. This means the computational cost of the ROS is 38 times that of the NOS. The average cross-sectional area of stories 1–3 of the ROS is 1.3 times as large as that of the NOS, and the average cross-sectional area of stories 7–9 of the ROS is half that of the NOS. The Monte Carlo simulation is carried out for the ROS and NOS with samples of size 1,000, respectively. Table 11.4 shows the maximum value, the mean value, the standard deviation, the worst 10% value, the value of $Y_{2,38}(\mathbf{x})$ and the ratio of the number of samples smaller than $Y_{2,38}(\mathbf{x})$ of the observed maximum inter-story drift angles. The standard deviation in the ROS is 59% of that in the NOS. The results indicate that the use of several tens of samples enables us to predict the 10% worst value with good accuracy. The histogram and distribution in each story of the observed maximum inter-story drift angles are shown in Figures 11.5 and 11.6, respectively, which confirm that the ROS has a smaller variation of response than the NOS.

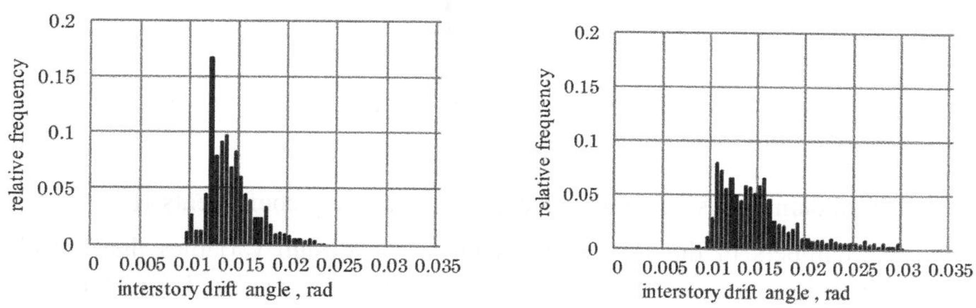

(a) ROS (b) NOS

Figure 11.5. *Histogram of observed maximum inter-story drift angles*

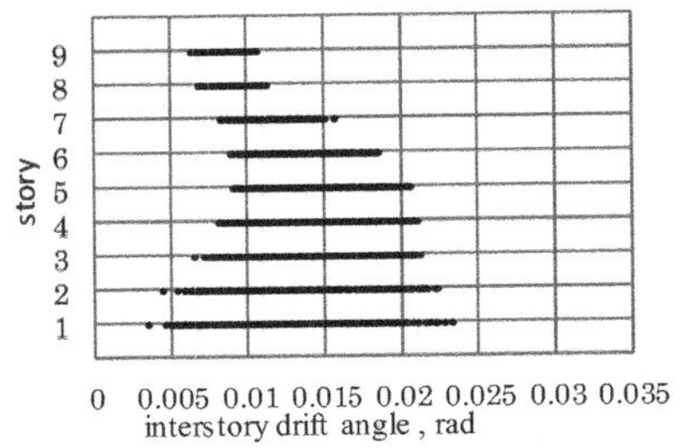

(a) ROS

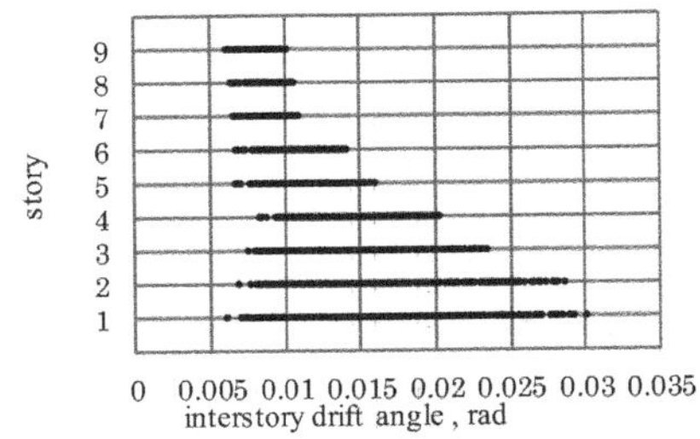

(b) NOS

Figure 11.6. *Observed maximum inter-story drift angles of each story*

11.4.3. *Multi-objective optimization considering robustness*

Equation [11.11] is solved for a 20-story shear frame with viscous dampers, as shown in Figure 11.7 (Ohsaki *et al.* 2019). Note that only the first and top stories are shown in Figure 11.7, and the intermediate stories have similar properties.

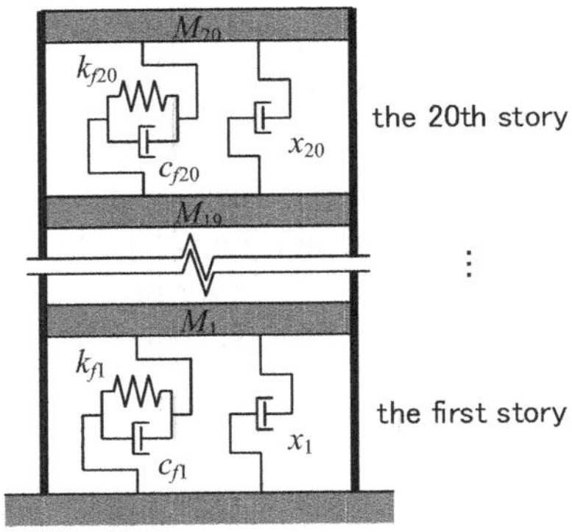

Figure 11.7. *A 20-story shear frame model (Ohsaki* et al. *2019)*

Structural damping is not considered, and the damping coefficients $(c_{f1},\cdots,c_{f20})$ of viscous dampers in the stories are taken as the design variables that are also denoted by the vector $\mathbf{x}=(x_1,\cdots,x_{20})$. Uncertainty is considered for story mass $(m_1,\cdots,m_{20})$, stiffness $(K_1,\cdots,K_{20})$ and the damping coefficient which is also a design variable. The nominal value of mass at each story is 4.0×10^5 kg, and the nominal values of story stiffness are proportional to the seismic shear coefficients defined by the Japanese building code. The fundamental natural period of the frame with nominal values of mass and stiffness is 1.6 seconds. Seismic responses are evaluated using the design displacement response spectrum $S_d(T,h)$, which is defined from the acceleration response spectrum as $S_d(T,h)=(2\pi/T)^2 S_a(T,h)$. The extended complete quadratic combination (CQC) method (Igusa *et al.* 1984; Yang *et al.* 1990) is used for the response evaluation of the structure with a non-proportional damping matrix due to the presence of viscous dampers in the stories. The maximum value among all the stories of the maximum inter-story drift is chosen as the representative response $g(\mathbf{x})$.

Since uncertainty is considered for the vectors of mass, stiffness and damping coefficient, which have 20 components, respectively, the vector $\boldsymbol{\theta}$ of uncertain parameters have 60 components in total as $\boldsymbol{\theta}=(\theta_1,\cdots,\theta_{60})\in\Omega$, where the uniform distribution in the range of $\pm10\%$ to the mean value is considered in each parameter as $\Omega=\{(\theta_1,\cdots,\theta_{60})\,|\,-0.1\le\theta_i\le0.1;\ i=1,\cdots,60\}$, which is multiplied by the mean value and added to the nominal value as

$$m_i^*=m_i+\theta_i\hat{m}\ ,\ K_i^*=K_i+\theta_{20+i}\hat{K}\ ,\ c_i^*=c_i+\theta_{40+i}\hat{c}\ \ (i=1,\cdots,20) \qquad [11.15]$$

where $\hat{m}$, $\hat{K}$ and $\hat{c}$ are the mean values of m_i, K_i and c_i, respectively. Eigenvalue analysis is carried out for each solution with 60 random parameter values, in order to obtain the natural periods and eigenmodes, and compute the representative response using the extended CQC method.

The [11.11] problem for this design problem is formulated using $n=150$, $\tilde{k}_i=1,\cdots,20$ and $\alpha=0.9$. Therefore, as shown in Table 11.2, the robustness level γ decreases from 1.000 to 0.113 as k is increased from 1 to 20:

$$\left.\begin{array}{ll}\underset{\mathbf{x}}{\text{Minimize}} & Y_{1,150}(\mathbf{x}),\ Y_{2,150}(\mathbf{x}),\ \cdots,\ Y_{20,150}(\mathbf{x})\\ \text{subject to} & x_i\ge0,\ \sum_{i=1}^{20}x_i\le\overline{c}\end{array}\right\}$$

where $\overline{c}$ is the upper bound for the total value of damping coefficients. Pareto optimal solutions are found using non-dominated sorting genetic algorithm II (NSGA-II) (Deb 2001) available in the global optimization toolbox of Matlab R2018a. The population size is 200, the number of generations is 117 and the elitist strategy is used. Transient analysis is carried out 150 times for each solution by generating 150 different sets of uncertain parameter values with uniform distribution. This means $200\times150\times117$ function evaluations are performed in total. The nominal solution is also found by assigning the nominal values to all uncertain parameters.

As a result of optimization, 70 different Pareto optimal solutions are found. Among them, the values of $Y_{1,150},\cdots,Y_{20,150}$ of three typical Pareto optimal solutions A, B and C are plotted in Figure 11.8(a) and (b). Solution A has the smallest value for $k=1$ among all the 70 Pareto optimal solutions; solution B has the smallest value for $k=16,18,19$; and solution C for $k=2,\ldots,6$. Solution C has the minimum variance of $Y_{1,150},\cdots,Y_{20,150}$ among the 70 solutions. Note that there are nine solutions, among the Pareto optimal solutions, that minimize $Y_{k,150}$ for different values of *k* in 1, …, 20. Solutions A and B have small $Y_{k,n}(\mathbf{x})$ values for small k ; however, the decrease of $Y_{k,n}(\mathbf{x})$ for larger k is very small. By contrast, for solution C, $Y_{k,n}(\mathbf{x})$ has a large value if k is small; however, it rapidly decreases as k is increased. The values of $Y_{1,150},\cdots,Y_{20,150}$ are also plotted for the nominal solution in Figure 11.8(a) by assigning random parameter values. As seen from the figure, the nominal solution is far from optimal, and obviously the responses of all orders have large values.

The values of design variables representing the damping coefficients $(c_{f1},\cdots,c_{f20})$ of the dampers in the stories are plotted in Figure 11.9(a) for Pareto optimal solutions A, B and C, in order to show that they have similar distributions with a small difference in the middle stories. Optimal additional damping coefficients of solution A and the nominal solution are compared in Figure 11.9(b), which shows that the nominal solution has smaller/larger damping ratios in the upper/lower stories than the Pareto optimal solution. The Monte Carlo simulation is also carried out for the solutions with samples of size 10,000, respectively. Table 11.5 shows the maximum value, the mean value and the standard deviation of the observed maximum inter-story drifts. The standard deviation in solutions A, B and C is about 40% of that in the nominal solution. The results indicate the effectiveness of the multi-objective approach.

Figure 11.8. *Relation between the maximum inter-story drift and order* k *for the nominal solution and three Pareto optimal solutions: (a) a plot including the nominal solution; (b) a detailed plot of Pareto optimal solutions (Ohsaki* et al. *2019)*

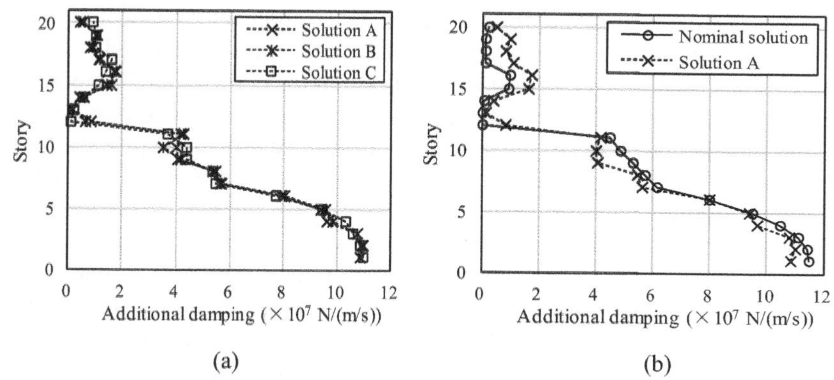

Figure 11.9. *Distributions of additional damping coefficients: (a) Pareto optimal solutions A, B and C; (b) solution A and the nominal solution (Ohsaki* et al. *2019)*

	Max (mm)	Mean (mm)	Std. Dev. (mm)
Nominal solution	17.0	12.9	0.61
Solution A	14.8	12.7	0.24
Solution B	14.9	12.7	0.24
Solution C	13.7	12.7	0.22

Table 11.5. *Summary of observed maximum inter-story drifts*

11.5. Conclusion

A method has been presented for the worst-case analysis of structures based on a random sampling approach and order statistics. By using this method, we can predict the approximate worst value with pre-assigned accuracy based on a clear definition of the worse values using the distribution-free tolerance interval of order statistics. Furthermore, a multi-objective optimization formulation for robust design has been developed based on the fact that the kth maximum value of the representative response corresponds to an approximate worst response with the robustness level γ for a specified confidence α among the samples with a specified number. Accordingly, the Pareto optimal solutions with various levels of robustness can be obtained by solving the multi-objective optimization problem of minimizing the representative responses with various orders.

To examine the applicability of the method when applied to practical problems, a robust optimum design of a nine-story building frame subjected to seismic excitations has been found by considering uncertainties in the ground properties and the phase of the design seismic motion. It has been shown that the robust design has a smaller standard deviation of the maximum inter-story drift angle than the nominal optimal design against the uniform random distribution of uncertain parameters. The application results indicate that several tens of samples can predict the approximately 10% worst value. Pareto optimal solutions have also been generated for a 20-story shear frame subjected to seismic motions. Uncertainty is considered in the design variables, which are the damping coefficients of the viscous dampers in the stories, the story mass and the story stiffness. The results suggest that the optimal damping coefficients depend on the order of response representing the level of robustness to be specified by the designer.

11.6. References

Ben-Tal, A., El Ghaoui, L., Nemirovski, A. (2009). *Robust Optimization*. Princeton University Press, Princeton, NJ.

David, H.A. and Nagaraja, H.N. (2003). *Order Statistics*. John Wiley & Sons, New York.

Deb, K. (2001). *Multi-Objective Optimization Using Evolutionary Algorithms*. John Wiley & Sons, New York.

Elishakoff, I. and Ohsaki, M. (2010). *Optimization and Anti-Optimization of Structures Under Uncertainty*. Imperial College Press, London.

Hardin, B.O. and Drnevich, V.P. (2002). Shear modulus and damping in soils: Design equations and curves. *Geotech. Spec. Publ.*, 98(118 II), 1459–1484.

Igusa, T., Der Kiureghian, A., Sackman, J.L. (1984). Modal decomposition method for stationary response of non-classically damped systems. *Earthq. Eng. Struct. Dyn.*, 12(1), 121–136.

Iyengar, R.N. and Rao, P.N. (1979). Generation of spectrum compatible accelerograms. *Earthq. Eng. Struct. Dyn.*, 7(3), 253–263.

Jekel, C.F. and Haftka, R.T. (2020). Risk allocation for design optimization with unidentified statistical distributions. *AIAA Scitech 2020 Forum*, 6–10 January, Orlando, FL.

Krishnamoorthy, K. and Mathew, T. (2009). *Statistical Tolerance Regions: Theory, Applications, and Computation*. John Wiley & Sons, New York.

McKenna, F. (2011). OpenSees: A framework for earthquake engineering simulation. *Comput. Sci. Eng.*, 13(4), 58–66.

Ohsaki, M. and Katsura, M. (2012). A random sampling approach to worst-case design of structures. *Struct. Multidiscip. Optim.*, 46(1), 27–39.

Ohsaki, M., Yamakawa, M., Fan, W., Li, Z. (2019). An order statistics approach to multiobjective structural optimization considering robustness and confidence of responses. *Mech. Res. Commun.*, 97, 33–38.

Schnabel, P.B., Lysmer, J., Seed, H.B. (1972). SHAKE: A computer program for earthquake response analysis of horizontally layered sites. Report, Pacific Earthquake Engineering Research Center, University of Berkeley, CA.

Yamakawa, M. and Ohsaki, M. (2013). Worst-case design of structures using stopping rules in k-adaptive random sampling approach. *Proc. 10th World Congr. Struct. Multidiscip. Optim* [Online]. Available at: https://mae.ufl.edu/mdo/Papers/5314.pdf.

Yamakawa, M., Nihei, M., Tatibana, M., Hukazawa, K., Ohsaki, M., Nakamura, T., Yamanouchi, H. (2015). Modeling and simulation of spring steel damper based on parameter identification with a heuristic optimization approach. *J. Mech. Sci. Technol.*, 29(4), 1465–1472.

Yamakawa, M., Ohsaki, M., Watanabe, K. (2018). Robust design optimization of moment-resisting steel frame against uncertainties of surface ground properties using order statistics. *Japan–China Workshop on Analysis and Optimization of Large-Scale Structures*. Structural Engineering of Buildings Laboratory, Department of Architecture and Architectural Engineering, Kyoto University [Online]. Available at: http://hdl.handle.net/2433/231246.

Yang, J.N., Sarkani, S., Long, F.X. (1990). A response spectrum approach for seismic analysis of nonclassically damped structures. *Eng. Struct.*, 12(3), 173–184.

Zhigljavsky, A. and Žilinskas, A. (2008). *Stochastic Global Optimization. Stochastic Global Optimization*. Springer US, New York.

List of Authors

Haym BENAROYA
Rutgers University
USA

Noël CHALLAMEL
University of Southern Brittany
France

Matthias FAES
KU Leuven
Belgium

Giovanni FALSONE
University of Messina
Italy

Maurice IMHOLZ
KU Leuven
Belgium

Nan JIANG
Beihang University
China

Julius KAPLUNOV
Keele University
UK

Benedikt KRIEGESMANN
Hamburg University of Technology
Germany

Rossella LAUDANI
University of Messina
Italy

Maurice LEMAIRE
Sigma Clermont
France

Giora MAYMON
Rafael Ltd.
(Retired)
Israel

David MOENS
KU Leuven
Belgium

Bowen NI
Beihang University
China

Makoto OHSAKI
Kyoto University
Japan

Zhiping QIU
Beihang University
China

Roberta SANTORO
University of Messina
Italy

Izuru TAKEWAKI
Kyoto University
Japan

Dirk VANDEPITTE
KU Leuven
Belgium

Lei WANG
Beihang University
China

Xiaojun WANG
Beihang University
China

Walter V. WEDIG
University of Karlsruhe
Germany

Makoto YAMAKAWA
Tokyo University of Science
Japan

Index

A, B, C

D, E, F

G, I, K

L, M, N

O, P, Q

R, S, T

U, V, W

Summary of Volume 1

Chapter 6. Buckling and Post-Buckling of Parabolic Arches with Local Damage

Uğurcan EROĞLU, Giuseppe RUTA, Achille PAOLONE and Ekrem TÜFEKCI

Chapter 7. Inelastic Microbuckling of Composites by Wave-Buckling Analogy

Rivka GILAT and Jacob ABOUDI

Chapter 8. Quasi-Bifurcation of Discrete Systems with Unstable Post-Critical Behavior under Impulsive Loads

Mariano P. AMEIJEIRAS and Luis A. GODOY

Summary of Volume 2

Chapter 5. On the Geometric Nonlinearities in the Dynamics of a Planar Timoshenko Beam

Stefano LENCI and Giuseppe REGA

Chapter 6. Statics, Dynamics, Buckling and Aeroelastic Stability of Planar Cellular Beams

Angelo LUONGO

Chapter 10. Dynamic Soil Stiffness of Foundations Supported by Layered Half-Space

Yang ZHOU and Wei-Chau XIE

Other titles from

iSTE

in

Mechanical Engineering and Solid Mechanics

2021

DAHOO Pierre Richard, POUGNET Philippe, EL HAMI Abdelkhalak
Applications and Metrology at Nanometer Scale 1: Smart Materials, Electromagnetic Waves and Uncertainties
(Reliability of Multiphysical Systems Set – Volume 9)

2020

LEDOUX Michel, EL HAMI ABDELKHALAK
Heat Transfer 1: Conduction

SALENÇON Jean
Elastoplastic Modeling

2019

BAYLE Franck
Reliability of Maintained Systems Subjected to Wear Failure Mechanisms: Theory and Applications
(Reliability of Multiphysical Systems Set – Volume 8)

BEN KAHLA Rabeb, BARKAOUI Abdelwahed, MERZOUKI Tarek
Finite Element Method and Medical Imaging Techniques in Bone Biomechanics
(Mathematical and Mechanical Engineering Set – Volume 8)

IONESCU Ioan R., QUEYREAU Sylvain, PICU Catalin R., SALMAN Oguz Umut
Mechanics and Physics of Solids at Micro- and Nano-Scales

LE VAN Anh, BOUZIDI Rabah
Lagrangian Mechanics: An Advanced Analytical Approach

MICHELITSCH Thomas, PÉREZ RIASCOS Alejandro, COLLET Bernard, NOWAKOWSKI Andrzej, NICOLLEAU Franck
Fractional Dynamics on Networks and Lattices

SALENÇON Jean
Viscoelastic Modeling for Structural Analysis

VÉNIZÉLOS Georges, EL HAMI Abdelkhalak
Movement Equations 5: Dynamics of a Set of Solids
(Non-deformable Solid Mechanics Set – Volume 5)

2018

BOREL Michel, VÉNIZÉLOS Georges
Movement Equations 4: Equilibriums and Small Movements
(Non-deformable Solid Mechanics Set – Volume 4)

FROSSARD Etienne
Granular Geomaterials Dissipative Mechanics: Theory and Applications in Civil Engineering

RADI Bouchaib, EL HAMI Abdelkhalak
Advanced Numerical Methods with Matlab® 1: Function Approximation and System Resolution
(Mathematical and Mechanical Engineering SET – Volume 6)
Advanced Numerical Methods with Matlab® 2: Resolution of Nonlinear, Differential and Partial Differential Equations
(Mathematical and Mechanical Engineering SET – Volume 7)

SALENÇON Jean
Virtual Work Approach to Mechanical Modeling

2017

BOREL Michel, VÉNIZÉLOS Georges
Movement Equations 2: Mathematical and Methodological Supplements
(Non-deformable Solid Mechanics Set – Volume 2)
Movement Equations 3: Dynamics and Fundamental Principle
(Non-deformable Solid Mechanics Set – Volume 3)

BOUVET Christophe
Mechanics of Aeronautical Solids, Materials and Structures
Mechanics of Aeronautical Composite Materials

BRANCHERIE Delphine, FEISSEL Pierre, BOUVIER Salima,
IBRAHIMBEGOVIC Adnan
From Microstructure Investigations to Multiscale Modeling:
Bridging the Gap

CHEBEL-MORELLO Brigitte, NICOD Jean-Marc, VARNIER Christophe
From Prognostics and Health Systems Management to Predictive
Maintenance 2: Knowledge, Traceability and Decision
(Reliability of Multiphysical Systems Set – Volume 7)

EL HAMI Abdelkhalak, RADI Bouchaib
Dynamics of Large Structures and Inverse Problems
(*Mathematical and Mechanical Engineering Set – Volume 5*)
Fluid-Structure Interactions and Uncertainties: Ansys and Fluent Tools
(Reliability of Multiphysical Systems Set – Volume 6)

KHARMANDA Ghias, EL HAMI Abdelkhalak
Biomechanics: Optimization, Uncertainties and Reliability
(Reliability of Multiphysical Systems Set – Volume 5)

LEDOUX Michel, EL HAMI Abdelkhalak
Compressible Flow Propulsion and Digital Approaches in Fluid Mechanics
(Mathematical and Mechanical Engineering Set – Volume 4)
Fluid Mechanics: Analytical Methods
(Mathematical and Mechanical Engineering Set – Volume 3)

MORI Yvon
Mechanical Vibrations: Applications to Equipment

2016

BOREL Michel, VÉNIZÉLOS Georges
Movement Equations 1: Location, Kinematics and Kinetics
(Non-deformable Solid Mechanics Set – Volume 1)

BOYARD Nicolas
Heat Transfer in Polymer Composite Materials

CARDON Alain, ITMI Mhamed
New Autonomous Systems
(Reliability of Multiphysical Systems Set – Volume 1)

DAHOO Pierre Richard, POUGNET Philippe, EL HAMI Abdelkhalak
Nanometer-scale Defect Detection Using Polarized Light
(Reliability of Multiphysical Systems Set – Volume 2)

DE SAXCÉ Géry, VALLÉE Claude
Galilean Mechanics and Thermodynamics of Continua

DORMIEUX Luc, KONDO Djimédo
Micromechanics of Fracture and Damage
(Micromechanics Set – Volume 1)

EL HAMI Abdelkhalak, RADI Bouchaib
Stochastic Dynamics of Structures
(Mathematical and Mechanical Engineering Set – Volume 2)

GOURIVEAU Rafael, MEDJAHER Kamal, ZERHOUNI Noureddine
From Prognostics and Health Systems Management to Predictive Maintenance 1: Monitoring and Prognostics
(Reliability of Multiphysical Systems Set – Volume 4)

KHARMANDA Ghias, EL HAMI Abdelkhalak
Reliability in Biomechanics
(Reliability of Multiphysical Systems Set –Volume 3)

MOLIMARD Jérôme
Experimental Mechanics of Solids and Structures

RADI Bouchaib, EL HAMI Abdelkhalak
Material Forming Processes: Simulation, Drawing, Hydroforming and Additive Manufacturing
(Mathematical and Mechanical Engineering Set – Volume 1)

2015

KARLIČIĆ Danilo, MURMU Tony, ADHIKARI Sondipon, MCCARTHY Michael
Non-local Structural Mechanics

SAB Karam, LEBÉE Arthur
Homogenization of Heterogeneous Thin and Thick Plates

2014

ATANACKOVIC M. Teodor, PILIPOVIC Stevan, STANKOVIC Bogoljub, ZORICA Dusan
Fractional Calculus with Applications in Mechanics: Vibrations and Diffusion Processes
Fractional Calculus with Applications in Mechanics: Wave Propagation, Impact and Variational Principles

CIBLAC Thierry, MOREL Jean-Claude
Sustainable Masonry: Stability and Behavior of Structures

ILANKO Sinniah, MONTERRUBIO Luis E., MOCHIDA Yusuke
The Rayleigh–Ritz Method for Structural Analysis

LALANNE Christian
Mechanical Vibration and Shock Analysis – 5-volume series – 3rd edition
Sinusoidal Vibration – Volume 1
Mechanical Shock – Volume 2
Random Vibration – Volume 3
Fatigue Damage – Volume 4
Specification Development – Volume 5

LEMAIRE Maurice
Uncertainty and Mechanics

2013

ADHIKARI Sondipon
Structural Dynamic Analysis with Generalized Damping Models: Analysis

ADHIKARI Sondipon
Structural Dynamic Analysis with Generalized Damping Models: Identification

BAILLY Patrice
Materials and Structures under Shock and Impact

BASTIEN Jérôme, BERNARDIN Frédéric, LAMARQUE Claude-Henri
Non-smooth Deterministic or Stochastic Discrete Dynamical Systems: Applications to Models with Friction or Impact

EL HAMI Abdelkhalak, RADI Bouchaib
Uncertainty and Optimization in Structural Mechanics

KIRILLOV Oleg N., PELINOVSKY Dmitry E.
Nonlinear Physical Systems: Spectral Analysis, Stability and Bifurcations

LUONGO Angelo, ZULLI Daniele
Mathematical Models of Beams and Cables

SALENÇON Jean
Yield Design

2012

DAVIM J. Paulo
Mechanical Engineering Education

DUPEUX Michel, BRACCINI Muriel
Mechanics of Solid Interfaces

ELISHAKOFF Isaac *et al.*
Carbon Nanotubes and Nanosensors: Vibration, Buckling and Ballistic Impact

GRÉDIAC Michel, HILD François
Full-Field Measurements and Identification in Solid Mechanics

GROUS Ammar
Fracture Mechanics – 3-volume series
Analysis of Reliability and Quality Control – Volume 1
Applied Reliability – Volume 2
Applied Quality Control – Volume 3

RECHO Naman
Fracture Mechanics and Crack Growth

2011

KRYSINSKI Tomasz, MALBURET François
Mechanical Instability

SOUSTELLE Michel
An Introduction to Chemical Kinetics

2010

BREITKOPF Piotr, FILOMENO COELHO Rajan
Multidisciplinary Design Optimization in Computational Mechanics

DAVIM J. Paulo
Biotribolgy

PAULTRE Patrick
Dynamics of Structures

SOUSTELLE Michel
Handbook of Heterogenous Kinetics

2009

BERLIOZ Alain, TROMPETTE Philippe
Solid Mechanics using the Finite Element Method

LEMAIRE Maurice
Structural Reliability

2007

GIRARD Alain, ROY Nicolas
Structural Dynamics in Industry

GUINEBRETIÈRE René
X-ray Diffraction by Polycrystalline Materials

KRYSINSKI Tomasz, MALBURET François
Mechanical Vibrations

KUNDU Tribikram
Advanced Ultrasonic Methods for Material and Structure Inspection

SIH George C. *et al.*
Particle and Continuum Aspects of Mesomechanics